Fort Hall

ll south from the Snake River
miles to a line from Wells, Nevada on the west
Utah border, a distance of 70 miles."

Oregon Cattle Trail and Emigrant Trail

1870's—Louis Harrell came to Idaho with first herd
of cattle for Uncle Jasper Harrell imported from Texas.

1872—1882—Jasper Harrell developed Shoe Sole Ranch in
three watersheds, Humboldt, Snake, and Great Salt Lake.

Goose Creek

California Trail

Oakley

1872—J.E. Bower, foreman for Jasper Harrell
discovered summer ranch on Trout, Goose, and Big Creek.

MOUNTAINS
Range
ad Line
Ridge
RANCHO
RANDE

IDAHO
UTAH

1886—Sparks-Tinnin purchased Joseph
Armstrong holdings, including Winecup brand.

NECUP
IELD

DEAD LINE RIDGE—Line established by
John Sparks beyond which, to the west and south,
range sheep bands were not welcome.

Grouse Creek

Central Pacific Railroad
CPRR
To Promontory Point
Completed 1868
Merged in 1885 with the
Southern Pacific.

Promontory
Mts.

1881—John Sparks and John Tinnin
purchased Jasper Harrell's ranches
on Thousand Springs Creek.

Thousand Springs Creek
Bonneville Basin

Great

Salt

Lake

GAMBLE
RANCH

Checkerboard Area Railroad Grant
Leased to John Sparks 1885

ntello

BONNEVILLE SALT FLATS
Sparks—Tinnin
Winter Range

ot
ak

Land mark of Hasting's Cutoff
on California Trail.

DD Fickert © 1984

arks (50%), Jasper Harrell (49%), A.J. Harrell (0.8%)
2%) incorporate Sparks—Harrell Company in California
scribed value of $1,000,000.

J. A. Y.

B. A. S.

CATTLE
IN THE
COLD
DESERT

James A. Young
B. Abbott Sparks

UTAH STATE
UNIVERSITY
PRESS
Logan, Utah
84322-9515

Library of Congress Cataloging in Publication Data

Young, James A. (James Albert), 1937–
    Cattle in the cold desert.

    Bibliography: p.
    Includes index.
    1. Beef cattle—Great Basin—History—19th century.
2. Ranchers—Great Basin—History—19th century.
3. Ranchers—Great Basin—Biography. 4. Ranch life—
Great Basin—History—19th century. 5. Range ecology—
Great Basin—History—19th century. 6. Sparks, John,
1843–1908. 7. Harrell, Jasper, 1830–1901. 8. Grazing—
Environmental aspects—Great Basin—History—19th
century. 9. Great Basin—History—19th century.
I. Sparks, B. Abbott, 1919–   II. Title.
SF196.U5Y68   1985   636.2′00979   85–22529
ISBN 0-87421-123-9

CREDITS

The authors wish to acknowledge and thank the following sources for photographs appearing in the book.

James A. Young, plates 1-17, 21, 23, 28, 30, 31, 32, 34, 35; also, cover, prologue, section pages I, II.

Nevada Historical Society, plates 18, 25, 26, 27, 29, 33, 36, 39-48; also, section pages III, IV.

Northeastern Nevada Museum, plates 20, 38.

U.S. Geological Survey, plates 22, 24.

Norman Glaser Collection, plate 19.

Tulare County Historical Society, plate 37.

The authors wish to thank the following for their courtesy in granting permission to reprint from copyrighted material:

Western Historical Quarterly: From "Haymaking: The Mechanical Revolution on the Western Range," 14 (July 1983): 311-26.

Don Fickert: Endsheet map of John Sparks's empire.

J.A. Young and R.A. Evans, eds.: Chart "Sagebrush Communities Distributed on Alluvial Fans." From "The Physical, Biological, and Cultural Resources of the Fund Research and Demonstration Ranch, Nevada," USDA/SEA, *Agricultural Reviews and Manuals*, ARM-W-11, June, 1980.

I. C. Russell: Boundary of the Great Basin. From *Geological History of Lake Lahontan*, (Washington, D.C.: U.S. Geological Survey, 1885).

To the Memory
of
John Sparks, Cattleman,
and to those who followed
on to the New Lands

CONTENTS

ILLUSTRATIONS

FIGURES

TABLE

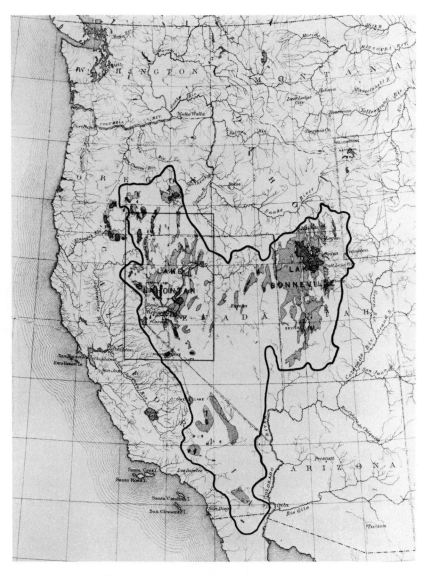

Figure 1. Boundary of the Great Basin. From I. C. Russell, *Geological History of Lake Lahontan*, U.S. Geological Survey, 1885.

ACKNOWLEDGMENTS

JAMES A. YOUNG

I greatly appreciate the many hours Glenda Eskstrom and Jamie Peer spent typing the many drafts of this manuscript. Appreciation is also expressed to Guy Rocha, Nevada State Archivist, and Phil Earl, Nevada Historical Society, who were always ready to help track down details and sources. The interlibrary loan staff of the Life and Health Sciences and Getchell libraries, University of Nevada, spent many hours obtaining material used in this manuscript. I am equally appreciative of the many other librarians and individuals who provided answers to my numerous requests.

Several individuals with the Agricultural Research Service of the U.S. Department of Agriculture encouraged and prompted the development of this manuscript. They include Dr. Raymond A. Evans, Research Leader, Dr. Edward B. Knipling, former Area Director, and Dr. Peter H. Van Schaik, former Associate Area Director. Howard Sherman contributed both his encouragement and his skill as an editor to this project.

Newton and Andy Harrell of Twin Falls, Idaho, provided a valuable link between the present and past historical characters such as Henry Harris. The firsthand accounts of these two cowboys sparkle with the flavor and color of the glory days of the Sparks-Harrell ranching empire.

I am very grateful to Dr. Charles S. Peterson, Utah State University, for believing in the potential contribution of *Cattle in the Cold Desert* to western history. Special thanks to Alexa West and Linda Speth, Utah State University Press, for taking a chance on a different kind of history manuscript.

Most of all, I thank my wife Cheryl and children Theresa, Patrick, and Nancy Yini for their patience and understanding.

B. ABBOTT SPARKS

My sincere appreciation to Jim Young for his invitation to join him in the completion of this work, which he conceived and researched through all essential stages. He and his associates were a pleasure to work with.

Phil Earl and Guy Rocha were also of invaluable assistance. Leta Lassetter, my assistant for ten years, devoted a year to the compilation of background data to assist me in the development of this manuscript. Don Fickert and Laura Jacobus made significant contributions to graphics and copy editing.

Russell E. Bidlack, editor of the *Sparks Quarterly*, Mrs. Grey Golden, former Texas State Archivist, and Mary Sparks Matthews added valuable information, as did Leland and Pat Sparks of San Francisco, California, and Nancy Sparks Lawrence, of Twin Falls, Idaho. Other Sparks family background was contributed by B. Abbott Sparks, Sr. and Van J. Sparks, of Pauls Valley, Oklahoma, both great-nephews of John Sparks, as well as Estha Scoggins of Georgetown, Texas. The many personal contributions to the completion of this work were both inspiring and indispensable. I join Dr. Young in his special tribute to Alexa West, Linda Speth, and Dr. Charles Peterson.

INTRODUCTION

ON CATTLE AND COLD DESERTS

All things occur within a larger context. Behind the individual events or circumstances is the grand mosaic of surrounding events—the geologic history, the human history, the environment. Thus, around the people and land of the Great Basin, there is a hint, a flavor, of events which have greater significance when taken as a whole rather than as individual events or circumstances. Against the backdrop of the cold desert's sagebrush/grasslands is the pageant of man and his herds. It is easy from a twentieth-century perspective to try to discern the underlying "meaning of things," in both a scientific and historical sense, but to have recognized those specific events as meaningful when they were underway would tax anyone's intuition and intellect. Thus, it is doubtful if those early pioneers, ranchers, and sheepmen had a perspective about what were significant events as they were actually happening. This difficulty is perfectly understandable, since even today we have no perspective when events are actually happening, when their place in the larger context has not yet become clear.

However, as we try to point out in this book, if we are attentive, we can begin to gain a very useful sense of the larger context even while the events are in progress. We can do this by sensitizing our awareness to fragments of history and science that we frequently ignore. These fragments are like puzzle pieces that, when recognized and assembled, give us an understanding of a greater meaning of things, a larger context.

Thus, we have taken the disciplines of science and history with their individual events and circumstances and placed them in a specific place for a specific time to see them woven together into a larger view of the Great Basin. The chapters alternate between land and man; but, in a sense, they are inseparable, and that is the main thrust of this volume—the larger picture of man and his cattle in the cold desert.

Another purpose of this volume is to provide perspective on the influence of grazing animals on the ecology of the sagebrush/grasslands. The scope of the coverage is designed to provide background information for those who earn their livelihood from grazing animals on the sagebrush/grasslands, for professionals who manage such lands, and individuals living in or outside the sagebrush/grasslands environment who are interested in the quality of life in this ecosystem.

This volume traces the history of man and his herds and flocks of domestic animals as they exploited the forage resources of a pristine environment. It is not the entire story, only the introduction and expansion period from roughly 1860 through the end of the nineteenth century. Essentially, this volume consists of a discussion of scientific principles and philosophies set in the context of historical events. This book is not meant to be a biography of the principal characters such as John Sparks or Jasper Harrell; however, it uses the life experiences of these pioneers to illustrate how man, as the herdsman, interacted with the sagebrush/grasslands environment.

The story of the exploitation of the grazing resources of the sagebrush/grasslands is a rerun of what has happened to virtually the entire surface of the earth, except for the most bleak arctic wastes and dense tropical jungles. The difference is that the sagebrush/grasslands was one of the last great vegetation resources to be suddenly, radically, and irrevocably changed by the introduction of domestic livestock. Because it was a relatively recent event, the process of the development of the livestock industry in the sagebrush/grasslands can be reconstructed with relative clarity.

Environmental quality is very much in the eyes of the beholder. A major objective of this account is to provide perspective for decisions on the nature of environmental quality in the sagebrush/grasslands. It is most difficult for one to perceive one's own homeland being degraded. This is especially difficult if the rate of environmental change is nearly static, with fathers relating to sons that the rangelands have not changed. Environmental changes in the sagebrush/grasslands tend to reflect sudden catastrophic events followed by long struggles toward equilibrium in a fluctuating environment.

To provide the reader with a sense of urgency regarding such environmental catastrophic events, we introduce this volume with a prologue, an account of five violent days in August 1964. This event is well after the time period (1860-1900) on which we concentrate, but the events of that August were so dramatic that for a brief instant, a portion of the general public questioned environmental quality in the sagebrush/grasslands and related it to events such as the introduction of large numbers of domestic cattle almost a century earlier.

Halfway through this volume, we relate another environmental catastrophe—the hard winter of 1889–90. Both events provided brief windows in time when man, as the herdsman of domestic livestock, was forced to look at the consequences of his actions. For the general public, such perceptions are fleetingly brief and easily lost in the haze of pseudo-stability that seems to follow sagebrush catastrophes.

The vegetation of the pristine sagebrush/grasslands was rather simple and extraordinarily susceptible to disturbance. The potential of the environment to support plant and animal life was limited by lack of moisture and often by accumulations of salts in the soil. The native vegetation lacked the resilience, depth, and plasticity to cope with concentrations of large herbivores. The plant communities did not bend or adapt; they shattered. This tends to make the review of grazing in the sagebrush/grasslands a horror story, resplendent in examples of what should not have been done. In perspective, the development of ranching in the sagebrush/grasslands was a grand experiment initiated by men willing to venture beyond the limits of accepted environmental potential to settle the Great Sandy Desert between the Rocky, Sierra Nevada, and Cascade mountains.

The dynamics initiated by the introduction of domestic livestock into the sagebrush/grasslands are still underway. Each plant community that composes an effective environment or habitat type in the sagebrush/grasslands is a biological measure of the interacting factors that make up the environment. The plant communities also mirror a century of disturbance. Interpretation of the influences of history on potential is essential to develop management schemes for the environment. There are no more pristine sagebrush/grasslands to exploit. We must learn to constructively use and restore the existing resource or sink into an endless spiral of further degradation.

PROLOGUE

PROLOGUE

FIVE DAYS IN AUGUST

At 10:00 A.M. on the Saturday morning of August 15, 1964, Elko, Nevada, was enjoying a pleasant summer morning with a few cumulus clouds drifting lazily east toward the towering bulk of the Ruby Mountains. The Ruby Mountains still supported patches of snow in sheltered, north-facing glacial cirques. Elko was hot and dry. Tourists felt sure they were going to die while driving across the baking Nevada desert. Highway 40, the major transcontinental route, passed through Elko and followed the Humboldt River Valley southwest to the sink where the river terminated near the Carson Desert.

This was the pioneers' route in the great rush to California in the 1850s.[1] The modern travelers welcomed the comfortable motels and hotels of Elko after the heat of the desert. The electric excitement of the casino made the tourist forget the endless sagebrush valleys and mountains. Liquor was cheap and the Elko night life was out of proportion to the size of the town, a transportation and supply center for the surrounding livestock ranches. Divided by two sets of transcontinental railway tracks, Elko County was advertised as the leading range livestock county in the nation. Some of the houses on the east side of the tracks had red lights over their doors. Cross-country diesel tractors and semitrailers parked in front with the motors idling. In early morning when the night life quieted down, the exhausted tourists dropped into bed only to be bounced out by the vibrations of a long freight train highballing through town.

Elko was a western town. The doctor, banker, barber, and undertaker dressed in western-cut suits or Levi's, and high-heeled cowboy boots. The plush restaurants and casinos were decorated in western motifs. Elko was a town with a cowboy heritage.

The characteristics of Elko County ranchers had changed greatly since the end of World War II. Beef and wool enjoyed a seller's market after price controls were lifted. According to an article in *Life* magazine, the only thing in short supply in Elko in the late 1940s was one-hundred-dollar bills.[2]

Many of the old ranchers sold to outside investors after the war.

Bing Crosby bought a ranch on the North Fork, and Jimmy Stewart bought the old H-D Ranch on the Thousand Springs Creek when the giant Utah Construction Company was sold. The H-D was the jewel in John Sparks's nineteenth-century ranching empire. Ranches became popular tax shelters for the new rich in the postwar era. Cattle prices remained relatively low during the early sixties, and very few ranchers were making a valid return of their capital investment.

As that August Saturday progressed, the Weather Bureau alerted the local fire suppression agencies, the United States Forest Service (USFS) of the U. S. Department of Agriculture (USDA), the Bureau of Land Management (BLM) of the U. S. Department of Interior (USDI), and the Nevada Division of Forestry (NDF) of a 40 percent chance of thunderstorms.[3] The BLM had the most land to worry about as 67 percent of the land area of Nevada was under its control.

Bob Carroll, the BLM Fire Control Officer in the Elko District Office, was worried. He had recently been successful in getting his fire crew increased from seventeen to twenty-seven men. He had started the summer with no standby fire crew. On June 27, three days after Carroll finally got his standby crew, they had their first fire. His fire crew was a collection of boys just out of high school, college students, and a few Indians from Duck Valley and the South Fork of the Humboldt. What they lacked in experience, they made up in enthusiasm. They were soon to get experience. By late July they had fought seventy-eight fires. This was equal to the total number of wildfires for 1963, and double the number for some previous years.

Range managers had realized for thirty years that much of the lower elevation sagebrush/grasslands were severely overgrazed. There were millions of sagebrush-dominated acres with virtually no herbaceous understory. The previous summer a BLM official from the Nevada headquarters in Reno told a USDA range scientist that these vast acreages of degraded sagebrush range were an asset rather than a liability. He claimed these degraded communities represented an environment frozen in time. They were fireproof because they lacked herbaceous vegetation to carry the fire. As the BLM developed plans and obtained funds, it believed these degraded communities could selectively be transformed into grasslands by plowing and seeding exotic (i.e., deliberately introduced) perennial crested wheatgrasses.[4]

The spring of 1964 was prolonged and wet. The alien (i.e., accidentally introduced) weed, cheatgrass, responded dynamically to the late spring moisture and produced a tremendous crop of herbage. On the northwest side of Emigrant Pass in July, 1964, range scientists clipped 4,000 pounds of cheatgrass per acre from experimental plots. On normal years, cheatgrass often produces only a few hundred pounds of herbage per acre and on dry years, virtually nothing. Cheatgrass is an annual which dries in early summer and provides a finely textured flash fuel that carries wildfire through sagebrush stands. The extent to which this accumulation of cheatgrass constituted a fire hazard is shown in the nine large fires that occurred during July. One of these fires burned 45,000 acres.

From the first to the fifteenth of August, there had been lightning storms somewhere in Nevada nearly every day. On July 29 there were flash floods at Yerington, Nevada, which tumbled cars on the highway like rubber toys. On August 11, Tonopah, Nevada, had a similar storm. Elko expected at least one intense lightning storm each year. All the fire-suppression officials could hope for was that the lightning storm and peak fire conditions did not coincide.

Bob Carroll could only wait and worry. His boss, Clair Whitlock, the Elko BLM District Manager, had left on vacation the night before and Bob had the lonely feeling he was sitting on a powder keg with a lit fuse. He checked the light plane he planned to fly for spotting lightning strikes, talked to the two Torpedo Bomber-Medium (TBM) pilots the BLM had on contract at the Elko airport. These pilots were flying aging World War II naval bombers modified for dumping a slurry of bentonite (a form of clay) on fires. Pilots had to love flying and danger to be involved in this business. Some years there were no fires and they spent a lot of time waiting around hot airports, tinkering with worn-out airplanes. If the pilot was lucky, some fire agency paid him a standby fee.

The standby fire crew in Elko viewed the 40 percent prospect of lightning with mixed emotions. The pumpers were serviced, the bull-dozers were loaded on the low-boy trailers and ready to go. The summer fire crew members welcomed fires because they brought relief from boredom and petty jobs around the BLM yards. Most of all, it meant overtime and extra dollars. On the other hand, it was Saturday night and the pleasures of Elko were awaiting their youthful enthusiasm. Working hard and playing hard were part of being on fire crews.

There had been so many fires during July that the professional range managers in the BLM had been forced to drop their regular jobs of administering grazing on the National Resource Lands to concentrate on fire suppression. Fire suppression was a physical thing that could be attacked and defeated through hard work and planning. Good land management was a nebulous thing with shifting goals, endless red tape, insufficient funds, and anything from apathy to outright hostility from ranchers.

Neither Bob Carroll nor the pilots and standby crews knew it, but their fate was sealed by 10:00 A.M. that Saturday morning. An upper level low-pressure system moved across central California in the early morning and at 10:00 A.M., it was located south of Lake Tahoe over the Sierra Nevadas.

At 1:00 P.M. there were towering cumulus and cumulonimbus clouds across the northern Great Basin. Winnemucca, Nevada, reported lightning and a sprinkling of rain in the early afternoon. Teleprinters clanked to life in Weather Bureau offices and fire radios crackled with the ominous news that the cloud base was 6,000 to 9,000 feet high, indicating a lack of moisture in the clouds. From Winnemucca to Wendover, Nevada, dry lightning cracked in sheets and massive single bolts on the tinderbox rangelands. The dry lightning flashed without accompanying rain.

By late Saturday afternoon about thirty-five fires had been reported. One experienced BLM fireman started for the Willow Creek fire in a pickup, and on the way he put out an estimated 300 acres of spot fires that no one had reported. The TMBs flew into the thunderstorms at incredible risk, and helped knock down sixteen of the fires. The remaining fires burned together to form the Boulder, Maggie Creek, Willow Creek, Palisade, and New Corral fire complex with the Upper Clover as a single fire. Someone coined the word "firestorm." It fit the situation and it stuck.

Bob Carroll luckily reached the District Manager, Clair Whitlock, at his first vacation stop. He told Clair he had to return to Elko; they had a firestorm. As the District Manager put the phone down to face his family with the news that the vacation was aborted, he wondered what the hell a firestorm was.

Considering the massive acreage burning, the suppression crew did quite well on Sunday with the limited manpower and equipment available. Ranchers organized their own crews to fight spot fires and generally cooperated with the government agencies in controlling the range fires. They were concerned about cattle in the fire areas and the areas of dry forage they planned to graze that fall being burned. The citizens of Elko were less concerned. It was the peak of the tourist season, hay was being cut on the ranches, and construction projects were in full swing. Anyone who wanted to work had a job and no one was interested in helping the government fight fires. "Let it burn, it will improve the range," was the oft-repeated comment in street-corner conversations.

Rancher Joe Peretti, of the 7 Lazy Y Ranch near Elko, was frantic about his cattle. Normally cattle were quite capable of getting out of the way of wildfires, but no one had experienced a firestorm on sagebrush rangelands. Lurid tales appeared in the Reno and Salt Lake papers suggesting that hundreds or even thousands of cattle had been killed and loss of fall forage would force early sales and millions in losses.

It was a long, sleepless Saturday night for the BLM personnel. Their problem on Sunday was to find bodies to press into fire crews. The state BLM office in Reno responded quickly with help from other districts. The Humboldt District of BLM took over the Kelly Creek fire that was near their boundary. The State Director of the BLM, J. Russell Penny, did not fit the stereotype of a successful bureaucrat. He was dynamic, could act decisively and organize effectively, but the only men immediately available from outside the bureau were winos off Lake Street, which at that time was the skidrow of Reno. They volunteered quickly enough, but their physical condition made them a safety hazard when they reached the fire lines.

By Sunday night, August 16, outside fire crews and overhead personnel were arriving in Elko and being dispatched. The overhead were supervisory personnel to take over the organization and direction of fire-suppression efforts. This was supposed to give Bob Carroll a chance for a few minutes rest, but he couldn't rest easily with field crews who had worked twenty-four hours straight without rest or food.

The Nevada Youth Training Center, a school for incorrigible boys committed by the courts, had dispatched its fifty-man crew under NDF foresters. During the 1930s and the days of the Civilian Conservation Corps (CCC), it was found that tough kids could be molded into fire crews if they were given quality leadership, training, and esprit de corps. The practice may have had no lasting reform influence on the boys; but at the time, dispatchers referred to fifty *men* from the Youth Training Center.

The BLM fire boss at the Elko headquarters had accumulated a reserve crew by midday on Monday. The crew was fed, loaded on buses, and going out the gate of the BLM yards when word was received of a large new fire near Elko. This was the Sherman fire and the buses were diverted to attack the blaze. About one-half hour later, the Grindstone fire was reported. Clair Whitlock was later to comment that the Sherman and Grindstone fires broke the back of the fire-suppression program. Not only did these two new fires exhaust all reserves, but the conditions which caused these two sleeper fires to explode spelled disaster. The temperature was rising, the humidity was dropping, and the afternoon winds were increasing.

The fire-fighting air force had been greatly augmented by Monday. The fire dispatcher allotted seven tankers to hit the Sherman fire when it was first reported. The fire was only seven miles from the airport, but despite the advantage of a short ferrying distance, the planes could not suppress the Sherman fire. The rate of spread was beyond the experience of any of the fire fighters, as an estimated 10,000 acres burned in two hours.

As the sun went down Monday night, there was a glow west of Elko. The particularly aromatic smell of sagebrush smoke was evident even in the air-conditioned casinos. The attitude of the townspeople underwent a sudden change. The BLM was besieged with volunteers. An army of fire fighters from throughout the West were pouring into Elko. George Zappettini, the Nevada State Forester, called on Governor Grant Sawyer for help. The Nevada National Guard was called out to assist in fire suppression, crew feeding, and transportation. The Nevada Air National Guard took aerial photographs of the burned area. Senators Howard Cannon and Alan Bible called on the Secretary of Interior, Stewart Udall, who promised to render all possible aid. Air Force bases Stead and Hill responded with medical teams, transportation, and more bulldozers.

Communications was a serious problem. There was not enough radio equipment, and much of what they had failed to function. Crews left Elko with drivers who had only vague ideas about where they were to deliver the fire fighters. A cook on the Boulder fire asked the fire boss to radio for 600 rations for dinner; later, he raised it to 1,000, then 2,000.

Communications problems haunted the fire fighters. The Boulder fire boss had fifteen aerial tankers dropping slurry in crossing patterns, with only one communications channel to direct all of them.

Wildfires in sagebrush vegetation normally explode during the late afternoon and quiet down after sunset when the winds die down. On Monday night, August 17, 1964, temperatures stayed relatively warm. The humidity was low, and the dawn came with swirling winds. The Boulder fire camp had to be moved several times during the night as the winds continually changed direction and rolled flames over the camps.

Tuesday was the day of the air force. Over forty aircraft were involved, including lead planes, tankers, and reconnaissance. There were twenty-one air tankers flying at one time. In addition, there was a large fleet of charter and Air Force planes. The runway was too short for the Air Force's largest planes, but C-119 cargo planes were unloaded as they rolled without stopping. The Federal Aviation Administration (FAA) set up a special visual flight rules control tower on top of the airport's terminal to handle the traffic. On Tuesday, August 18, Elko's airport handled more flights than Los Angeles International. Something relatively new in fire fighting was the use of fifteen helicopters, in which crews were transported to remote locations with speed and ease. Fire crews remember the relief they felt when a helicopter brought fifteen gallons of badly needed drinking water. The fire air force became supertechnical with a USFS research plane carrying infrared sensing equipment capable of determiming the extent of the fire through smoke. Another research plane from the Desert Research Institute, University of Nevada, seeded thunderheads with silver iodide crystals.

The TBM pilots were making a dollar per minute. They earned their money by coming in low through smoke and fire turbulence at ninety knots, which was below the cruising speed of their aging aircrafts. A converted B-17 showed up at Elko with a 1,000 gallon tank in its bomb bay. On its first run the entire load hit a concentrated area and dug a trench across the fire line. Depending on the size of the load, each slurry drop cost $300 to $500. On Tuesday, 79,000 gallons of retardant were dropped on the fires. One BLM fire observer's plane had a defective starter; he risked life and limb to turn the prop over by hand.

9

On Sunday, August 16, at 8:00 A.M. phones were ringing in BLM, National Park Service, and Bureau of Indian Affairs offices throughout the Southwest. The word was out—Elko, Nevada, had a huge fire and the Southwest Indian fire crews were needed. To build fire lines, fire bosses needed tough individuals who were used to the outdoors, and familiar with axes and shovels. The chronic unemployment of southwestern Indian reservations was familiar to southwestern land managers, and the Indians had the qualities that were required of fire fighters. But, the experiment also involved the Indian crews gaining acceptance. Every experienced fire boss in the West could relate problems he had had in picking up local, unemployed Indians and trying to make fire fighters out of them. Elko, more or less, fit the stereotype of western towns in its collective attitude toward Native Americans. Individually, the townspeople had many Indian friends and there were many successful Indian laborers, ranch hands, and professionals in the Elko area. This was success measured in terms of Anglo standards. On many of the larger ranches, local Indians were a significant part of the labor force and the haying crews. Although there were no signs in the plush Elko restaurants and casinos that said, "Indians not welcome," those Indians who patronized the bars that faced the railroad tracks were not welcome uptown.

The Southwest Indian crews had proven themselves as fire fighters on hundreds of fires. On the deadly Haystack burn in northern California, they had held Indian Creek Ridge with hand tools when heavy equipment operators turned and ran.[5] Local papers often devoted more space to how the Indians danced for rain than to how skillfully they constructed and held fire lines.

Harold "Pete" Davis of Socorro, New Mexico, was a liaison officer with a twenty-five-man Taos Indian crew. The crew chief, Albert Martinez, rousted his men out of the Taos pueblo early Sunday morning on August 16. Edward Archulelata said goodbye to his wife, Cesarita and five children, and left with the crew to fight fires in Elko County, Nevada, for $2.09 per hour. Air Force C-119 planes airlifted the Taos crew to Elko. Buses took them from the airport to a downtown tourist restaurant where they were fed a steak dinner. The waitress and townspeople in the restaurant were very glad to see them. After the meal, they were taken to the Elko High School football field, which was being used as a staging area, and then transported to the Palisade fire with a crew of Zia Indians.

The Navaho Number 1 crew from Rock Point Trading Post at Chinle, Arizona, had a more exciting experience getting to the fires. The crew was composed of eighteen-year-olds on their first fire and men in their late fifties who had toured the West fighting fires. When they arrived at the fire, no one knew where the fire lines were. They also lacked sufficient hand tools. After many delays, paper sleeping bags were distributed. There were not enough bags for the entire group but they bedded down in the dust as best they could to await the dawn. Twice during the night the camp was awakened with cries that the fire was coming closer and they had to move in a mad scramble. For sheer terror, a fast-moving wildfire approaching at 30 MPH in the middle of the night in strange country would be hard to top.

The Indian crews were not the only ones on the move at the Boulder fire on that terror-filled Sunday night. The Air Force had sent tractors equipped with bulldozers to the fire with Airmen 3C operators fresh out of heavy equipment school. They had operated tractors through obstacle courses on training grounds, but not in rugged terrain in the dark with the danger of being trapped by fast-moving fires. The fire boss put experienced Indian fire fighters on the tractors beside the airmen to offer advice and steady taut nerves. Some of the Southwest Indians had limited English vocabularies, but they calmed the nerves of the young Air Force men with a look, a nod of the head, or a hand on the shoulder.

Experienced wildfire fighters knew the characteristics of sagebrush fires and tried to use this knowledge to their advantage. Fires burn uphill much better than downhill. Experienced fire fighters never get upslope from advancing fires, especially if there are rocks or cliffs that block their escape. Near Orvada, Nevada, there is a monument to the memory of a crew of CCC men who failed to obey these rules. However, on the Boulder fire the flames refused to obey the rules. Flames roared upslope and crested the ridges. Then, instead of dying down as is prescribed in the standard script, the flames rolled down the back slope in one fell swoop. Fire fighters were astounded to see flames advance against the wind through some trick of convection.

The convection column above the Boulder fire was so towering and

11

so intensely hot that it literally broke up thunderheads. The blazing walls of flames distilled gases from the sagebrush fuel ahead of the fire itself. Occasionally, these gases would explosively ignite causing the flames to leapfrog ahead in a spectacular fashion. The Brewer's sparrow is a bird species characteristic of degraded sagebrush rangelands. On the Boulder fire, fire crew members reported seeing Brewer's sparrows being kicked up from the sagebrush cover by advancing flames. The birds tried to fly away from the advancing fire, but the suction created by the updraft impeded their progress until the zone of distilling gases caught up with the birds. When the distillation zone ignited, the birds vanished in a puff of incandescent gases.[6]

Ross Ferris, the fire boss on the Maggie Creek fire, cussed the whirlwinds on Monday afternoon. I.C. Russell described the whirlwinds that start out on salt flat playas and dance along as sensuous columns that seemed to support the entire summer Nevada sky.[7] The whirlwinds that danced across the Maggie Creek fire were dirty with ashes and dust. They picked up cow chips and pieces of sagebrush bark, some still smoldering. When these embers hit unburned fuel, the fire was again off and running.

The area of the Maggie Creek fire was visited by the botanist, P.B. Kennedy in 1901.[8] The plant communities of the Maggie Creek watershed were largely degraded at that time, but the alien cheatgrass had not yet invaded the area. Now, sixty-three years later, cheatgrass provided the fuel that flashed the fire from shrub to shrub.

Marion Escobar had come from the Las Vegas office of the BLM to be fire boss on the Palisade fire. He was pleased with the Zia and Taos Indian crews who built fire trails so fast that fire bosses had to stay out of their way. Despite their maximum efforts, the new fire raced over the hills at 35 MPH and eventually consumed 30,000 acres. That night Pete Davis's crew of Taos Indians were sitting in the dust eating dinner, when a young Hot Shot Crew from the Boise Fire Control Center marched through the Indians shouting cadence and stirring the dust. The Indians turned to Davis to ask a collective, "Why?"[9]

Residents of Elko were worried about the effect the fires would have on the upcoming hunting seasons. Each fall thousands of hunters, many of them from out of state, descended on Elko County to harvest upland game birds and mule deer. It was big business for Elko.

On Tuesday, the army of fire fighters grew to over 3,000 men, 260 vehicles, 60 tractors, 15 helicopters, and 40 fixed-wing aircraft. Fifteen federal, state, and county agencies were cooperating. The fires appeared contained, but Nature was not yet through. Tuesday afternoon brought one of the strangest phenomena of the entire fire complex. It was observed by many, but no clear explanation ever emerged. At 4:00 P.M. Tuesday afternoon the winds across the Boulder fire suddenly intensified in a rolling cloud of dust, ash, and smoke flicked with flames. From the Maggie Creek fire it looked like the end of the world was rolling down on the crews at 60 MPH to join with the Boulder in one huge fire. Witnesses say that a "funnel cloud" hit the Boulder fire, after which confusion reigned. Crews were busy avoiding the flames, while trying to see with eyes that were stinging from smoke, dust, and ashes. Smoke hid the fire from the planes circling overhead.

Fire headquarters in Elko received frantic phone calls from ranchers at Tuscarora, Nevada, that a new fire had broken out to the north of the Boulder fire. Crews were dispatched to reinforce a scratch force of ranchers. When the winds hit at 4:00 P.M., the Boulder fire made a spectacular run of five miles in ten minutes! There was a chilling aspect to this. By chance, there was no one in its way. No one could have run or driven over rough roads fast enough to escape.

The fires were contained by Wednesday. The crews then started picking up the trash and cleaning up the mess of 3,000 men who had camped in the sagebrush. There were strong words expressed by northeastern Nevada ranchers and townspeople that something was wrong with the sagebrush environment. Why were firestorms rushing across the landscape? It had to be the government's fault. The old-timers said it was never like this in their youth. There were many postmortem discussions among professionals. As one Elko fire boss laconically put it, "If the darn fire had lasted five weeks instead of five days, we would have gotten command and communication problems solved."

All the forces involved mounted a dedicated response to the 1964 firestorms in Elko County. This amounted to emergency treatment of the symptoms without a postmortem search for the cause. Cattle had grazed in Elko County in large numbers for about ninety years before the 1964 firestorm.[10] During that period much of the pristine environment of the sagebrush/grasslands of western North America was irrevocably altered. What we know as ranching, or the cowboy culture, evolved. Thick, juicy steaks, one of the products of ranching, evolved as the symbol of luxury in American diets.

Was this agricultural development built on consumptive exploitation of the natural resources of the sagebrush/grasslands? Will the twenty-first century result in the reduction of Nevada's sagebrush/grasslands to annual grass and weed ranges? Biologically, conversion to annuals ranks somewhere between feasible and probable. It can happen and probably will unless, as a culture, we collectively change our ways. Wildfires such as the 300,000-acre Elko fire complex in 1964 are the triggering mechanisms that set sagebrush/grasslands ecosystems into dynamic successional changes. The flames of the wildfires destroy the shrubs that have frozen the degraded plant communities into sagebrush dominance and allow the herbaceous vegetation to respond. The wildfires, such as the 1964 firestorms, are not the cause of environment degradation; they are the product. Thus, the degradation of sagebrush rangelands is not the product of the activities of one resource consumer or of the land-management agencies; it is a product of our culture.

Ranching as we now know it was not transplanted to the sagebrush/grasslands; it evolved in place. Environmental Impact Statements are now written on the impact of grazing on National Resource Lands. Without historical perspective as to how the sagebrush/grasslands arrived at the current ecologic, social, and economic conditions that exist, it is impossible to accurately interpret Environmental Impact Statements.

Once the ashes cooled on the 1964 Elko fires, land management professionals initiated a crash program to try to rehabilitate the burned areas. But eventually, townspeople, ranchers, and many professional land managers forgot their critical need to find out what conditions in the sagebrush environment brought fires to the edge of Elko. The problems have not gone away. The accidental combination of forage fuel and dry lightning will return. Through application of appropriate technology, the environment can be restored to an approximation of the pristine environment with stable communities of forbs, shrubs, and grasses. If the burned sagebrush ranges are not restored, the alien weeds will inherit the sagebrush/grasslands. The way is thus paved for repeated burnings and a continuing downward spiral of degradation.[11]

14

Notes

1. Dale Morgan, *The Humboldt Highroad of the West*, (New York: Farrar & Rinehart, Inc., 1943).

2. Roger Butterfield, "Elko County," *Life*, 18 April 1949, Vol. 76 (16): 98–114.

3. Unless otherwise cited, information is from testimony recorded at the fire critique held by Bureau of Land Management (BLM) in Elko, and on file in BLM state office, Reno, Nevada. See clippings from *Chicago Tribune, Los Angeles Times, San Francisco Chronicle, Deseret News* (Salt Lake City, Utah), and *Burlington Free Press* (Burlington, Vermont) for 16–20 August 1964.

4. Personal communication to James A. Young from Dr. Raymond A. Evans, Investigation Leader, Pasture and Range Management Project, U. S. Department of Agriculture, Agricultural Research Service, Reno, Nevada.

5. Personal experience of James A. Young, Siskiyou County, California, 1955.

6. *Reno Evening Gazette*, 20 August 1964, and BLM records.

7. I.C. Russell, *Geological History of Lake Lahontan*, (Washington, D.C.: Government Printing Office, U. S. Geological Survey, 1885). [Government Printing Office hereafter referred to as GPO.]

8. P.B. Kennedy, *Summer Ranges of Eastern Nevada Sheep*, (Reno, Nevada: Nev. Agric. Exp. Sta. Bull. 55, University of Nevada, Reno, Nevada, 1903).

9. Hot Shot Crews were organized by various fire suppression agencies, given special training and equipment, and used as shock troops to knock down fires as soon as they were reported and before they got out of control. To add to their esprit de corps, these crews often gave themselves names, and stitched them on their Levi jackets.

10. Lester W. Mills, *A Sagebrush Saga*, (Springville, Utah: Art City Publishing Co., 1954).

11. J.A. Young, and R.A. Evans, "Population Dynamics after Wildfires in Sagebrush Grasslands," *J. Range Manage.*, (1978) 31:283–89. Much of the 300,000 acres that burned in the 1964 fire was restored by a superb revegetation effort conceived by J. Russell Penny, directed by Bill Malencik. For details, see Guy R. Sheeter, "Secondary Succession and Range Improvement after Wildfires in Northeastern Nevada," M.S. thesis, University of Nevada, Reno, Nevada, 1968.

I
THE
OPEN LAND

THE OPEN LAND

There was not a single gate to open or fence to obstruct the movement of man and his herds between Salt Lake City at the foot of the Wasatch, and Genoa at the eastern base of the Sierra Nevada. In the valleys of the central Great Basin, a person could ride for one hundred miles and never find a tree with sufficient shade to protect a rider and horse from the glare of the noonday sun. The barren salt flats reflected the dazzle of light and created mirages on which the bases of the mountains seemed to float. The mountains were islands in a sea of desert aridity, straining to reach up to the clouds that yielded water, the gift of life in this environment. The land awaited man.

CHAPTER
1

GRAY OCEAN OF SAGEBRUSH

The relentless silver-gray reflection of the cold desert's sagebrush landscape could not help but impress those first few travelers who ventured West. The Oregon-bound travelers began to have their first taste of sagebrush near Fort Laramie, Wyoming, and as they proceeded westward, the gray ocean of sagebrush increased. John C. Fremont had difficulty getting a wagon through dense stands of sagebrush on the Snake River Plains. The perennial grasses that did exist in the sagebrush were mature, dry, and harsh when the Oregon settlers reached the sagebrush plains in early autumn.[1]

The endless uniformity of a sagebrush-dominated landscape tends to blur differences and hide the details of this land's plant communities. Sagebrush/grasslands are plant communities in which species of sagebrush form an overstory and various perennial grasses form the understory. In western North America, from southern Canada to northern Mexico, a group of closely related sagebrush species comprise an edemic (i.e., occurring here only) section of the worldwide genus *Artemisia**. They occur in varying amounts over 422,000 square miles in eleven western states. Sagebrush/grasslands occur in various types in all the western states but the discussion here will focus on those range plant communities that occur in the Intermountain area.

The Intermountain area is the vast region from the Rocky Mountains on the east to the Sierra Nevada and Cascade ranges on the west and bounded on the south by the true warm deserts and on the north by coniferous forests. The Intermountain area is a cold desert—a semiarid to arid region where the winters are bitterly cold and often snowy.

When settlers first entered this vast area, how did they view it? The term "contact period" is generally used by anthropologists to indicate

---

*Author's note: For readability, scientific names have been kept to a minimum. An equivalency table of common and scientific names for plants and animals precedes the Index.

the period of first contact between indigenous cultures and European or American trappers, explorers, and settlers. Sources of information for the contact period in western North America are largely trappers' journals often written or edited after the actual contact in the field. The major subject of these journals was the fur trade with comments on the general environment usually secondary. Comments on plant communities were frequently only asides and often must be interpreted from other statements. Since trappers tended to travel and camp in river or stream valleys where beaver were found, most of the written comments concern these areas. In the Great Basin, where the mountain ranges tend to run north and south and are oriented in echelon, travelers could avoid crossing the steep mountain ranges by going around the mountains. These nineteenth-century travelers tended to group all shrubs in the Intermountain area as sage or wormwood even though the trails passed through greasewood- or saltbush-dominated landscapes. Essentially, one can use the records left by trappers and early explorers of the Intermountain region to justify any preconceived ideas of the pristine vegetation of the sagebrush/grasslands.

The extensive records kept by the early Mormon colonists provide the best account of the pristine sagebrush/grasslands environment. George Stewart, then with the Intermountain Forest and Range Experiment Station, summarized the available records in 1940. Stewart was uniquely trained and experienced for the role. After first becoming a successful agronomist and college teacher, he joined the Forest Service as a range ecologist. Stewart's position in the Mormon church allowed him access to the records of the early church colonies. Interestingly enough, Stewart takes quotations from the journals written during the contact period to emphasize the abundance of grass under pristine conditions, while T.R. Vale concludes the opposite and stresses shrub dominance.[2] Generally, the records indicate that the abundance of shrubs increased as the Mormon settlers moved westward and southward in the Intermountain region until they passed into true desert vegetation.

Early travelers along the Oregon Trail had the opportunity to view a good cross section of the sagebrush ecosystem from Fort Laramie to the Columbia River. The travelers' opinion of the sagebrush country was partially dependent on the time of the year they crossed the Snake River Plains.[3] If they crossed the area in late summer or early fall, the journals stressed sagebrush, lack of forage, and dust. Because of the time interval required for an overland journey from Missouri to Oregon, most travelers crossed the sagebrush/grasslands during the late summer. These travelers crossed the sagebrush/grasslands after they had traveled through the Great Plains, one of the world's foremost grasslands. Thus, the travelers crossed the Plains during the peak of the growing season, and the sagebrush/grasslands obviously suffered by comparison.[4]

Most of the activities of the fur trappers were confined to the upper Snake River and adjacent areas. However, the hunt for beavers was much like the later prospecting for gold. The trappers followed up virtually every stream in search of wealth. Anglo-American exploration and fur trade began in the Snake River area in 1809 and continued until 1846.[5] Peter Skene Ogden, who commanded the Snake River Brigade for the Hudson's Bay Company, left detailed and believable records of his extensive travels in the Intermountain region.[6] The Snake River Brigade traveled with 200 horses for riding and the packing of their supplies, traps, and furs.[7] These animals depended on forage obtained along the line of march and in the vicinity of winter camps. The trappers who composed the brigades were the epitome of hunters. When the brigade traveled in the eastern Snake River country where there were American bison, the hunters would exasperate Ogden with their wanton killing of game animals and failure to pay attention to the business of trapping. However, when traveling in the Great Basin and south-central Idaho, Ogden considered forage for his horses easy to find, potable water scarce, and big game for food difficult for his professional hunters to locate.

Throughout much of western North America, the location of the initial settlements in the pristine environment was determined by the chance discovery of minerals or convenient points along transportation routes. The Mormon settlements of the Intermountain area are an exception to this rule. These were agricultural settlements that had to pick specific environments if they were to survive. Within thirteen years after 1847, when Salt Lake City was founded, a series of outlying colonies was established from Lemhi in Idaho to Genoa, Nevada, to San Bernardino, California. The sites for these settlements were carefully selected to include areas suitable for irrigation to support intensive agricultural and grazing lands capable of producing the meat, milk, and draft animals necessary for the colonizers to survive. The extensive records kept by these early colonies provide us with the best description of the environment available for the development of productive sagebrush/grasslands.

In the late eighteenth century, the American bison apparently occasionally extended its range across the northern portion of the sagebrush/grasslands into northeastern California, the Malheur and Harney basins of eastern Oregon, and even to the Columbia Basin. This is roughly the bluebunch wheatgrass portion of the sagebrush/grasslands. The American bison had withdrawn its range from northern Nevada long before historic times. Therefore, the Thurber's needlegrass portion of the sagebrush/grasslands had no concentration of large herbivores under pristine conditions. In the early nineteenth century the number of American bison on the upper Snake River and Green River drainages was first increased from hunting pressure east of the Rocky Mountains; and then, after 1830, the populations west of the mountains were exterminated. The spread of trade that provided rifles to the Indians hunting for robes, promiscuous hunting by trappers, and several severe winters contributed to the destruction of the American bison west of the Rocky Mountains.[8]

The buffalo or American bison was the only large herbivore to exist in large numbers on the sagebrush/grasslands during recent geologic times.[9] At the close of the Pleistocene, the upper Snake River Plains were grazed by native species of mastodon, camels, horses, and ancestors of the American bison. All of these animals, with the exception of the bison, became extinct. Rabbits, rodents, and harvester ants became the major consumers in the sagebrush/grasslands. Certainly the pronghorn remained, but as evidenced by the difficulty of the professional hunters of the fur brigades in getting meat, their populations were scarce in the Intermountain area and very scarce in the Great Basin. The Goshute Indians of Deep Creek in eastern Nevada practiced the communal activity of driving pronghorns with systems of traps, blinds, and barriers. Under pristine conditions the pronghorn populations in the various eastern Nevada valleys were sufficient to support one of these drives each decade. With this low a population, it is no wonder the trappers had a hard time killing camp meat.[10]

The pronghorn is a true native to North America while the large members of the deer family in North America are fairly recent emigrants from Asia by way of the Bering Strait Land Bridge.

Pronghorns are a true large herbivore of the sagebrush/grasslands. The females bear their young in May and the kids are soon following their mothers through the sagebrush. In summer bands of three to twenty animals are common, and in winter the pronghorns collect in even larger bands. The animals are migratory in the sense that those on the higher part of the summer range move down to lower territory where there is less snow in winter; but essentially, they always live in the sagebrush/grasslands.

The sagebrush/grasslands are also populated by rodents. Many of the rodent species are adapted to utilize metabolic water to satisfy their moisture requirements. This means that the rodents seldom if ever actually drink water; rather, they obtain their moisture requirements from the food they eat. This is a tremendous competitive advantage compared to large herbivores who are limited to grazing within the range of infrequent waterholes. A second adaptation of many of the rodent species is the use of underground burrows to modify the environmental extremes of the Great Basin.

The black-tailed jackrabbit (*Lepus californicus*) is probably the largest consumer of plant material in many parts of the sagebrush/grasslands. A traveler across Nevada sees more jackrabbits of this species than all the other small animals combined. The black-tailed jackrabbit is abundant in virtually all the lower elevation sagebrush areas. A black-tailed jack was spotted a mile out on the barren 14-Mile Salt Flat east of Fallon, Nevada, and one was killed at the 11,700 foot level on Mt. Jefferson in Oregon, illustrating the range of the species in the Intermountain area.[11]

The fluctuation in numbers of this species is so marked that almost every Nevada resident has noted the phenomenon. The cause of the sudden crashes in rabbit populations has been suggested to be tularemia, a disease caused by *Bacillus tularense*. This disease also attacks man. In jackrabbits, the mortality rate may reach 90 percent of populations. When the black-tailed jackrabbit populations are near their peak in a given area, they can be extremely destructive to crops, especially irrigated fields located in a general sagebrush environment. Under pristine conditions, the white-tailed jackrabbit (*Lepus townsendii*) was probably much more abundant than the black-tailed jackrabbit.

Reconstruction of the pristine environment is not limited to historical records. Range ecologists are continually looking for relic areas

23

where plant communities exist in equilibrium with the natural environment. These areas exist because of natural fencing, such as a mesa with sheer walls or lava flows that stock cannot cross. Relic areas can also exist because of slope angle or distance from stock water. In the sagebrush/grasslands water points are scarce and unevenly distributed across topography that is often rugged. This creates uneven utilization by grazing animals leaving areas long distances from water or on steep slopes ungrazed.

To look at this present sagebrush environment with an eye to differences in the past, one must first understand the complex vegetation structure behind this gray landscape. First, it is necessary to learn the identity of the major plant species, and then to recognize how the plant species fit together to form plant communities.

Within the Intermountain area there are two major subdivisions: the Snake River drainages and the Great Basin. The Snake River is part of the Columbia system which also contains the Columbia Basin that historically supported a northern extension of the sagebrush/-grasslands. In southern Idaho, the Snake River flows into a deep canyon walled by nearly vertical basalt cliffs. On top of these cliffs are extensive undulating plains that were formerly clothed with sagebrush. These plains are, rather obviously, known as the Snake River Plains.

Across a mountainous divide south of the Snake River Plains lies the other subdivision of the Intermountain area, the Great Basin. The Great Basin is a physiographic area having somewhat indefinite boundaries. Roughly, it lies between the Sierra Nevada on the west and the Wasatch Mountains on the east, but its tributary valleys extend over to Wyoming. To the southeast it grades into high plateaus near the Colorado River. Southward, the province extends through the Mojave Desert of California in Baja, California. Thus, the Great Basin comprises most of Nevada and Utah with a fringe in California, Oregon, Idaho, and Wyoming. The name, Great Basin, was first applied by John C. Fremont in 1844 when he scientifically established that no water drained into the ocean from this huge area.

The plant communities of the sagebrush/grasslands are a measure of the potential of the environment. (See Figure 1.) The history of the exploitation of the sagebrush/grasslands is also reflected in the present plant communities of the sagebrush/grasslands. The wall of flames

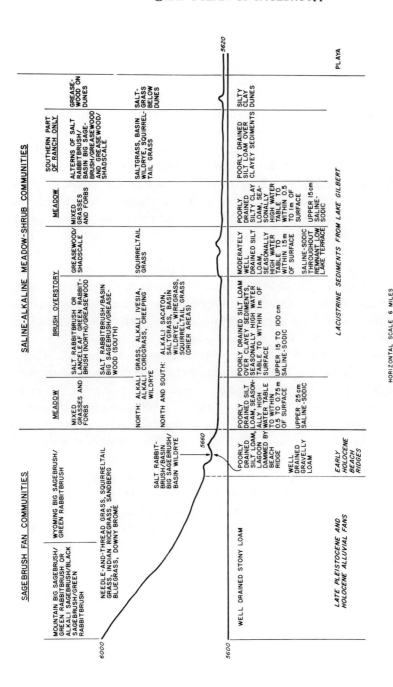

Figure 2. Sagebrush communities distributed on alluvial fans. From Young and Evans, 1980. Used with permission.

exploding through the big sagebrush/grasslands communities of Elko County in 1964 were a vivid expression of this history.

What were the environmental conditions that existed in the sagebrush/grasslands under pristine conditions? Beyond the curiosity factor of what the pristine environment was, the composition of plant communities in equilibrium with their environment serves as a benchmark to judge the quality of the rangelands. The composition of the plant community in equilibrium with its environment is judged to be in excellent *range condition*. As the plants that compose the pristine community change in abundance, with, for example the unpreferred shrubs increasing and the desirable perennial grasses decreasing, range condition drops to good, fair, or poor. The direction range condition is proceeding, either downward towards a degenerated condition or upward towards equilibrium, is called *range trend*. Range condition and trend are the basic parts of the science of evaluating the impact of grazing animals on the environment. Obviously knowledge of the pristine environment before domestic livestock were introduced is vital in establishing condition and trend standards.[12]

There are many kinds of sagebrush growing in the sagebrush/-grasslands, but the species that generally characterizes the environment is big sagebrush (*Artemisia tridentata*). The scientific name was first given to the species in 1841 by Thomas Nuttall to a collection from the plains of the Columbia River.[13] Big sagebrush is normally an erect shrub three to six feet tall, but occasionally is dwarfed. The trunk of the shrub is definitely woody with a stringy, fibrous bark. The entire plant presents a light gray-green appearance because of the silver-gray hairs on the leaves and new twigs. The color of the leaves reflects much of the sun's incoming radiation and protects the plants from desiccation in the sagebrush/grasslands' arid environment. The leaves are persistent through the winter and when the current year's growth begins in early spring, it is difficult to distinguish it from previous seasons' growth. Flowerheads appear in midsummer with flowering in late August and September. The yellowish flowers are borne in clusters on flower stalks. The brownish-black seeds begin to fall in October, and some persist until spring.[14]

There are three characteristics of big sagebrush that are especially significant to its ecology. First, this landscape-dominant shrub does not sprout when the aerial portion of the plant is burned in wildfires;

second, the species is composed of many ecologically distinct subspecies that in appearance or morphology are difficult to distinguish; and third, the essential oil content of the herbage of big sagebrush inhibits the growth of the rumen microflora in cattle and, to various degrees, other ruminants.

The fact that big sagebrush does not sprout after being burned in wildfires has fundamental significance in the ecology of the species. Fire and its role in the pristine environment of the sagebrush/grasslands is difficult to assess. In forests, the frequency of past fires can be reconstructed by examining fire scars that are left on the trunks of trees. These scars or "cat faces" record the frequency of fires in the annual rings of the trees by damaging the cambium layers.[15] In the sagebrush/-grasslands there are no trees to record the frequency of fires. The effect of wildfires in sagebrush/grasslands is to eliminate the landscape's dominant shrub. Under pristine conditions the net effect of wildfires was to release the native perennial grasses from competition. On steep north-facing slopes where good stands of perennial grasses persisted, the firestorm often did not burn the communities. The phenology or growth pattern of the native perennial grasses is much later than the alien annual cheatgrass. The fire season for pristine sagebrush/grasslands must have occurred in late August and early September versus the mid- to late-summer fire season with cheatgrass.

Once big sagebrush is consumed in a wildfire, the community that reoccupies the site is not devoid of shrubs. There are a number of shrubs that are subdominants to big sagebrush that have root or crown sprout after being burned. One of the important subdominant shrubs in big sagebrush communities is low rabbitbrush. This is a highly variable species with many distinct subspecies. Almost all of the subspecies and forms of low rabbitbrush are nonpreferred by large herbivores. Horsebrush sprouts from roots rather than dormant crown buds like low rabbitbrush. Not only is this shrub nonpreferred by browsers, it is also toxic. White-skinned animals are photosensitized by consuming herbage of horsebrush. There are several shrub species that occur occasionally in big sagebrush communities and sprout after being burned. Species of plum such as desert peach, green ephedra, and ribes have extensive underground stems or woody crowns called lignotubers that serve to protect buds and store food for regeneration after being burned. Green ephedra is an interesting species because it is a

27

Gymnosperm, more closely related to the pine trees than to the sagebrush species. Green ephedra does not have leaves in the commonly used sense, but green broomlike twigs. It is one of the few vividly green species in a gray environment.[16]

The net ecological effect of the big sagebrush being susceptible to wildfires is that the character of the landscape can be suddenly changed by a single catastrophic event. It takes from ten to fifteen years for big sagebrush to reinvade areas where it was destroyed by fire. If the fire hopped and skipped through the stand leaving many shrubs untouched, the rate of return is much faster. During the ten- to fifteen-year period after sagebrush burns, the sites are dominated by the root-sprouting shrubs.

How frequent were wildfires in the pristine sagebrush/grasslands? Henry Wright offered the unique analysis that the distance in time or interval between fires had to be greater than the time period of ten to fifteen years that the root sprouters dominate the communities after the sagebrush is burned. If this was not true, we would have had under pristine conditions the rabbitbrush/horsebrush/grasslands rather than the sagebrush/grasslands.[17]

Big sagebrush is the landscape-characterizing species of the sagebrush/grasslands, but within the species there are important subdivisions. To recognize the subdivision, it is necessary to carefully compare morphology or appearance, chemical differences as identified by thin layer chromatography, cytological evidence from the number, shape, and characteristic of the chromosomes, ecologic characteristics of the communities where the shrubs are found, and distributional patterns. From careful research studies the following subspecies have been recognized: 1) basin big sagebrush, 2) Wyoming big sagebrush, 3) mountain big sagebrush, and 4) subalpine big sagebrush. Generally mountain big sagebrush occurs at higher elevations, Wyoming big sagebrush occurs on drier alluvial fans, and basin big sagebrush on alluvial soils of the Great Basin, Snake River Plains, and innumerable mountain valleys.[18]

The third characteristic of big sagebrush that has great significance to the culturing of cattle in the sagebrush/grasslands is the essential oil content of the herbage. Sagebrush is very abundant, and the protein content of its herbage approaches or exceeds that of the cultivated alfalfa. The branchlets of sagebrush are not ridged or spiney to prevent

28

browsing. On the surface it would appear that sagebrush would be an excellent forage species. However, no vertebrate has evolved that is capable of breaking down and digesting highly lignified-cellulose plant material. The major herbivores, both wild and domesticated, capable of consuming and digesting coarse grasses are all ruminants. The rumen is a modified digestion system in which the animal provides anaerobic (oxygen-free) sites for the growth of microorganisms. The micro-organisms break down the woody plant material so it can be digested by the host animal. The end products of rumen bacterial fermentation, the volatile fatty acids, supply the major source of energy to the ruminant. When cattle consume small amounts of big sagebrush, there is no problem; but, if big sagebrush herbage constitutes a relatively large portion of the diet the activity of the rumen microflora is retarded or inhibited.[19] The same general trend is true for mule deer and elk, although these native big game animals can consume more sagebrush than cattle. The only native large herbivore that bases a large portion of its diet on big sagebrush is the pronghorn.

The inhibition of rumen microflora has been linked to the essential oil content of the sagebrush herbage. The term essential oil is a misnomer. They are not essential for the growth and reproduction of the plants. The compounds in sagebrush that inhibit rumen micro-organisms are volatile oils composed of terpene compounds and derivatives.[20] The amount of volatile oil that individual sagebrush plants contain depends on the season of the year, the site where the plant is growing, the environmental stress to which the plant is subjected, and the subspecies of big sagebrush involved. Mountain big sagebrush generally has a lower content of volatile oils than basin big sagebrush. Within subspecies, there are probably inherent differences in the quantity of volatile oils among strains of big sagebrush. This raises the possibility that plant breeders could produce a big sagebrush that does not inhibit the microflora of cattle and mule deer.[21]

The lack of large herbivores under pristine conditions in much of the sagebrush/grasslands and especially in the Great Basin is puzzling when compared to the essential oil content of big sagebrush herbage. Many large herbivores apparently base their forage selection on smell. Apparently the strong-smelling essential oil content of the herbage is of adaptive advantage to the shrubs in preventing browsing.[22]

Once big sagebrush plants become established, they occupy a site for a very long period. There is a poor correlation between the size of sagebrush plants and their age. Big sagebrush plants have been reported with over 200 annual growth rings. The huge, treelike plants of big sagebrush that are occasionally found along drainage ways growing on old meadow soils are seldom more than seventy or eighty years old. Their age probably reflects the time when the overgrazed meadows were desiccated by deepening channels, and brush species invaded.[23]

Just as all big sagebrush plants are not the same taxon, not all sagebrush is big sagebrush. In the western United States, there are approximately 226,370 square miles of big sagebrush out of the total of 422,200 square miles of sagebrush. The next most important species is silver sagebrush, 53,200 square miles; followed by black sagebrush, 43,300 square miles; and low sagebrush, 39,100 square miles. In the Humboldt Basin of northern Nevada, 45 percent of the landscape is dominated by species of sagebrush. Sagebrush species may be successional dominants in pinyon/juniper, mountain brush, and mountain conifer communities in addition to the 45 percent of the landscape that is true sagebrush/grasslands. Of this 45 percent of the landscape that is sagebrush, 40 percent is big sagebrush, and 5 percent low sagebrush. In northeastern California the sagebrush/grasslands are roughly 50 percent low and 50 percent big sagebrush.[24]

Silver sagebrush is often associated in the Intermountain area with the fine-textured soils and the seasonal flooding of old lake basins. Silver sage is restricted to the east side of the Sierra Nevada and eastern Oregon and does not extend down into the highly saline/alkaline soils in the depths of the Intermountain deserts. Silver sagebrush does sprout after the aerial portion of the shrub is removed; however, because of the nature of the habitat it occupies, it seldom is burned in wildfires. The various subspecies of silver sagebrush are widely represented in the northern Rocky Mountains and the major river valleys on the east slope of the Rocky Mountains.

Black and low sagebrush are the major components of the group of sagebrush species that are aptly named the "low sagebrush" type. These species are usually one-third the height of big sagebrush. Although similar in appearance, they often occupy contrasting landforms. Low sagebrush is usually on the oldest landform in a given landscape where soils have a well-developed clay horizon close to the surface. Because of this clay soil, this site takes up water very slowly and in the spring low sagebrush flats are extremely wet and sticky. Conversely, when dry, the soils are baked brick-hard.[25]

In comparison to basin big sagebrush, low sagebrush is highly preferred by big game and by domestic animals. However, when it comes to sheep, black sagebrush is a highly preferred winter-browse species. On the margins of the Carson Desert in western Nevada there are extensive stands of black sagebrush. After a century of winter use by sheep, these monospecific communities look like they were tended by a host of Louis XV gardeners. Each shrub is perfectly molded in ground-hugging exotic shapes.

Low sagebrush, as its name implies, is much reduced in stature compared to its cousin, big sagebrush. Often only a foot to eighteen inches high, low sagebrush spreads over the landscape in a tidy gray film compared to the uneven texture of big sagebrush landscapes. Because they are old landforms, low sagebrush flats often have biscuit and swale topography. The biscuits have shallow, mounded soil profiles and the swales may be devoid of soil with only stringers of frost-sorted rocks present. A plant community called "balds" occur on high, exposed ridges. These communities are dominated by low sagebrush shrubs scarcely four inches high and shaped by winter gales.

Throughout the range of the sagebrush/grasslands there are a series of plant communities delineated by the dominant shrub species and the understory grass species. To the unfamiliar, this network of species may appear to be a bewildering array of variability. However, the plant communities are repetitive and identifiable. To be able to recognize them is important because they are phytometers or living intergrading measurements of the effective environments of a given local ecosystem composed of soil, topography, climate, animals, and the plants themselves. The most well-known big sagebrush community is the big sagebrush/bluebunch wheatgrass, which predominates in the Columbia Basin, eastern Oregon, much of Idaho, and extends down the higher elevations into the Great Basin. This is a bunchgrass community in which the dominant grass occurs in a distinct bunch as opposed to a continuous carpet of grass. In a bunchgrass community, the ground is not covered with vegetation. Much of the soil surface is exposed, and only 20 to 30 percent of the area is covered with the canopies of plants.

To individuals with experiences in the humid East or the grasslands of the Great Plains, the sagebrush/grasslands with their bunchgrasses

31

appear poverty-stricken. However, the bunchgrasses are in equilibrium with their environment. Not only is precipitation limited in quantity throughout the sagebrush/grasslands, but the timing of the precipitation is largely out of phase with temperatures suitable for plant growth. Precipitation largely occurs as snow during the winter months. Plant growth occurs during the spring and is limited by cold temperatures in early spring and by the exhaustion of soil moisture in early summer. Into the brief spring growth period are crammed all the functions of growth and reproduction. Even in the semiarid grasslands of the Great Plains, moisture falls during the summer when temperatures are optimum for growth. On the Great Plains ten inches of spring-summer precipitation is much more effective than ten inches of winter precipitation in the Great Basin.

Bluebunch wheatgrass is just as much a landscape-characterizing species as big sagebrush. On steep slopes good stands of bluebunch wheatgrass resemble ranks of soldiers marching up the slope. Found hand-in-glove with bluebunch wheatgrass, but growing on more mesic situations, such as north-facing slopes, is Idaho fescue. At lower elevations, especially in the southern part of the Intermountain area, Thurber's needlegrass replaces bluebunch wheatgrass as the main understory species in big sagebrush communities. In virtually every big sagebrush community, the short-lived perennial grasses, squirreltail and Sandberg bluegrass, are found as minor components. Once the communities are disturbed these short-lived grasses rapidly respond to renew the community.[26]

There are several other species of perennial grass that form communities with sagebrush species. Communities of needle-and-thread grass often dominate areas with sandy soil. All of the needlegrass species are characterized by long awns on their seeds. The various species of needlegrass can be relatively easy to identify when they are in fruit by the shape and hairiness of the awns on their seeds. Needle-and-thread grass carries this to an extreme. The seed itself is needle sharp at its tip and the four-to-six-inch twisted awn is a coarse barbed thread.

Black and low sagebrush duplicate the big sagebrush communities in forming communities with the major perennial grasses. There is a low sagebrush/bluebunch wheatgrass community, low sagebrush/Idaho fescue, etc. and the same series exists for black sagebrush.

The perennial grasses begin growth in early spring. Most species flower in early summer and set seeds which are mature by midsummer. As they start growth in the spring the perennial grasses mobilize food, stored as carbohydrates, for energy to support their growth. If the grasses are heavily grazed during the early spring, year after year, without ever having a chance to rebuild their carbohydrate reserves, they will die. If the native perennial grasses are heavily grazed every spring, they never have a chance to produce seed for the establishment of new plants.

One thing that was missing from the sagebrush/grasslands was native annual grasses. Six-weeks fescue (*Vulpia octoflora*) is a diminutive native annual grass that briefly graces the sagebrush range in the spring, but it is not highly competitive. The herbaceous communities contain no vigorous, highly competitive annual grass capable of rapidly renewing the communities if they were destroyed by grazing animals. Many of the valleys of central and northwestern Utah that are in the big sagebrush zone had the aspect of true grasslands with dominance by bluebunch wheatgrass. Without question, various woody species of sagebrush, and especially big sagebrush, existed in all these environments. What is different in the present environment is the proportion of grasses versus shrubs.

Sagebrush communities are not usually colorful with wildflowers. Yet, the season after high-elevation sagebrush communities burn in wildfires, the renewed community can be a riot of color. After wildfires, large forbs like arrowleaf balsam root and mules ears, which are members of the sunflower family, have large showy flowerheads with bright yellow ray flowers. Mixing with the yellow of the daisylike coarse forbs are the delicate red shades of the Indian paintbrush flowers. Indian paintbrush roots often parasitize the roots of sagebrush plants, robbing them of nutrients. The deep-blue to purple colors of low larkspur spikes add radiant but deadly beauty to the gray-shrub landscape. The larkspurs are poisonous to livestock as is the death camus, a member of the lily family. Not all the flowers are robust or perennial. Under the shrubs occur the delicate flowers of Chinese house.

One flower picks a very specific habitat. The skunk monkey flower grows on the garbage dumps of harvester ant mounds. The harvester ants are important consumers in the sagebrush/grasslands. Virtually any oblique aerial photograph of a sagebrush community growing on an alluvial fan reveals a series of vegetation-free circles. These circles are the

33

mounds of the harvester ants. The ants strip all the vegetation from a six-foot-diameter circle and build a small soil mound some eight to ten inches in the center. The busy ants harvest seeds from various plants growing in the sagebrush communities. Some of the seeds have glands that the ants relish. The ants take the seeds into their underground chambers, remove the glands, and then discard the seeds which turn into a habitat for the skunk monkey flower.

The potential of the sagebrush/grasslands to support a cattle industry was something that had to be judged by mid-ninteenth-century agriculturalists before they risked capital and often their lives in introducing livestock to those areas. Essentially, the would-be sagebrush rancher had no precedent by which to judge the potential of the environment. The sagebrush/grasslands of the Intermountain region were a much more severe and undependable environment than had been previously encountered by Americans in their spread across the country.

The pristine environment that was available for exploitation by grazing consisted of a series of sagebrush/perennial bunchgrass communities. The species of sagebrush that formed the overstories in these communities were not preferred by cattle as browse and if eaten in quantity would cause the death of the browsing animals by disrupting their delicate symbiotic relationship with their rumen microflora.

In contrast to the Great Plains where great herds of American bison, pronghorn, and often deer and elk were displaced by the introduction of cattle, the sagebrush ranges of the Intermountain area had few large herbivores. Most importantly, the herbaceous vegetation that composed the forage resource of the sagebrush/grasslands was not preconditioned for heavy utilization by large herbivore herds.

By the mid-nineteenth century, the grand experiment was about to begin—the exploitation of the last virgin grazing resource in the United States—the grazing of the sagebrush/grasslands.

Notes

1. Comments on travel on the Oregon Trail. See John Townsend, *Narrative of a Journey Across the Rocky Mountains to the Columbia River*, pp. 107–639, in R. Thwaites, ed., *Early Western Travels*, Vol. 21, (Cleveland,

Ohio: Arthur Clark Company, 1906). See also F. Wisenzenius, *A Journey to the Rocky Mountains in the Year of 1839*, (St. Louis, Missouri: Missouri Historical Society, 1912); and, J.C. Fremont, *Report of the Exploring Expedition to the Rocky Mountains in the Year 1842, and To Oregon and Northern California in the Years 1843-44*, (Washington, D.C.: Gales and Seaton, 1845).

2. George Stewart, "History of Range Use," in *The Western Range*, Senate Document No. 199, 74th Congress, 2nd Session, (Washington, D.C.: GPO, 1936); and, T.R. Vale, "Presettlement Vegetation in Sagebrush/grassland Areas of the Intermountain West," *J. Range Manage.*, (1975) 28:32–36. For further information, see George Stewart, "Historic Records Bearing on Agricultural and Grazing Ecology in Utah," *J. Forestry*, (1941) 39:363–75. For the reconstruction of the vegetation of a specific valley, see A.C. Hull, and M.K. Hull, "Presettlement Vegetation of Cache Valley, Utah and Idaho," *J. Range Manage.*, (1974) 27:27–29.

3. "Comments on Travel on the Oregon Trail." [Uses the same sources as in Note #1 above.]

4. E.W. Tisdale, J. Hironaka, and M.A. Foshers, *The Sagebrush Region in Idaho*, (Moscow, Idaho: Agric. Expt. Sta. Bull. 512, University of Idaho, Moscow, Idaho, 1969).

5. Alexander Ross, *The Fur Hunters of the Far West*, M.M. Quaite, ed., (Chicago, Illinois: Lakeside Press, 1924).

6. P.S. Ogden, *Peter Skene Ogden's Snake Country Journals, 1824-1825 and 1825-1826*, E.E. Rich and H.M. Johnson, eds., (London: XIII Hudson's Bay Record Soc., 1950). For an overview of exploring the Great Basin, see G.G. Cline, *Exploring the Great Basin*, (Norman, Oklahoma: University of Oklahoma Press, 1963).

7. Personal communications to James A. Young from Mary Rusco, Nevada State Museum, Carson City, Nevada.

8. W.T. Hornaday, *The Extermination of the American Bison*, U. S. Natl. History Mus. Rept. for 1887, (Washington, D.C.: GPO, 1889). For additional information, see C.H. Merriam, "The Buffalo in Northeastern California," *J. Mammalogy*, (1926) 7:211–14.

9. B.R. Butler, "The Evolution of the Modern Sagebrush-grass Steppe Brome on the Eastern Snake River Plain," pp. 5–39, in Robert Elston, ed., *Holocene Environmental Change in the Great Basin*, (Reno, Nevada: Nevada Arch. Soc. Res. Paper No. 6, 1976).

10. M.H. Egan, and H.R. Egan, *Pioneering the West, 1846 to 1878*, W.M. Egan, ed., (Salt Lake City, Utah: Private Printing, 1917).

11. E. Hall, *Mammals of Nevada*, (Berkeley, California: University of California Press, 1946). See also, J.K. McAdoo, and J.A. Young, "Jackrabbits," *Rangelands*, 2 (1980): 135–138.

12. A.W. Sampson, "Succession as a Factor in Range Management," *J. Forestry*, (1917) 15:593–96.

13. H.M. Hall, and F.E. Clements, *The Phylogenetic Method in Taxonomy: The North American Species of Artemisia, Chrysothamnus and Atriplex*, (Washington, D.C.: Carnegie Inst., 1923). Original citation in Thomas Nuttall, *Trans. Amer. Phil. Soc.*, (1841) 2:398.

14. A.A. Beetle, *A Study of Sagebrush*, (Laramie, Wyoming: Wyoming Agric. Expt. Sta. Bull. 368, 1960).

15. S.F. Arno, and K.M. Sneck, "A Method for Determining Fire History in Coniferous Forest and Range," (Ogden, Utah: Expt. Sta., Forest Service, U. S. Dept. of Agric., 1977).

16. J.A. Young, and R.A. Evans, "Wildfires in Semiarid Grasslands," *Proc. XIII International Grassland Congress*, Leipzig, German Democratic Republic, 1977. See also J.A. Young, and R.A. Evans, "Population Dynamics of Green Rabbitbrush in Disturbed Big Sagebrush Communities," *J. Range Manage.*, (1974) 27:127–32.

17. H.A. Wright, L.L. Neuenschwander, and C.M. Britton, "The Role and Use of Fire in Sagebrush-grass and Pinyon Juniper Plant Communities," (Ogden, Utah: Expt. Sta., Forest Service, U. S. Dept. of Agric., 1978).

18. A.H. Winward, and E.W. Tisdale, *Taxonomy of the Artemisia tridentata Complex in Idaho*, (Moscow, Idaho: University of Idaho, College of Forestry, Bull. 19, 1977).

19. J.G. Nagy, H.W. Steinhoff, and G.M. Ward, "Effect of Essential Oils of Sagebrush on Deer Rumen Microbial Functions," *J. of Wildlife Manage.*, (1964) 28:784–90.

20. J.G. Nagy, "Wildlife Nutrition and the Sagebrush Ecosystem," pp. 144–68, *in Proc. Sagebrush Ecosystem: A Symposium*, (Logan, Utah: Utah State University, Logan, Utah, 1978).

21. Jeff Powell, "Site Factor Relationships with Volatile Oils in Big Sagebrush," *J. Range Manage.*, (1970) 23:42–46. See also D.L. Hanks, J.R. Brunner, D.R. Christensen, and A.P. Plummer, "Paper Chromatography for Determining Palatability Differences in Various Strains of Big Sagebrush," (Ogden, Utah: Expt. Sta., Res. Paper INT-101, Forest Service, U. S. Dept. of Agric., Ogden, Utah, 1971).

22. E.H. Cronin, Phil Ogden, J.A. Young, and William Laycock, "The Ecological Niches of Poisonous Plants in Range Communities," *J. Range Manage.*, (1978) 31:328–34.

23. C.W. Ferguson, and R.R. Humphrey, "Growth Rings of Sagebrush Reveal Rainfall Records," *Progressive Agriculture*, Summer Issue, 1959, p. 3.

24. A.A. Beetle, "Sagebrush." See also, Humboldt Basin statistics from U.S. Soil Conservation Service, "Water and Related Land Resources, Humboldt River Basin, Nevada," (Carson City, Nevada: U. S. Dept. of Agric., Report 12, 1966). Information on sagebrush in northeastern California from J.A. Young, R.E. Evans, and Jack Major, "Sagebrush Steppe," pp. 763–96 in M.G. Barbour and Jack Major, ed., *Terrestrial Vegetation of California*, (New York: John Wiley and Sons, 1977).

25. Ben Zamora, and P.T. Tueller, "*Artemisia arbuscula, A. longiloba*, and *A. nova* Habitat Types in Northern Nevada," *Great Basin Naturalist*, (1973) 33:225–42.

26. J.A. Young, et al., "Sagebrush Steppe." Also, J.F. Franklin, and C.T. Dryness, *Natural Vegetation of Oregon and Washington*, (Portland, Oregon: Pacific Northwest Forest and Range Expt. Sta., Tech. Rept. NW–8, Forest Service, U. S. Dept. of Agric., 1973).

CHAPTER 2

THE EXPLOITATION PAGEANT

The earliest rancher in the Intermountain area is a nebulous figure. It may have been a fur trapper or a freighter who first overwintered in this area and grazed his animals. Whoever that nebulous figure was, he participated in the first exploitation of the region's grazing resources.

The time period of 1860 to 1900 encompasses both the classical range operation period in which cattle were allowed to roam freely on unfenced native ranges with minimum husbandry, and also the period of ranch enterprise in which livestock production was stressed and in which some fences and conserved forage were employed in the culturing of livestock. Essentially, the evolution of the cattle ranch that evolved to exploit the sagebrush/grasslands is traced with ranches defined, in this instance, as being agricultural enterprises in which extensive acreages of native vegetation are utilized for the production of livestock, with red meat the product marketed.

Nebulous figures provide little interest in history books, so to illustrate the evolution of ranching in the sagebrush/grasslands, real cattlemen of history enter the story. Woven together with the story of the land of the Great Basin are the stories of Jasper "Barley" Harrell, his son, Andrew J. Harrell, and their long-time business associate John Sparks. To illustrate specific points, the activities of other ranchers and participants in this exploitation pageant will also be incorporated in the narrative.

The Harrells and Sparks were chosen for several reasons. The ranching enterprise they founded was one of the largest developed in the Intermountain area. It persisted as an operational unit until the mid-twentieth century, long after the death of the founders. Their ranching empire was centrally located in the Intermountain portion of the sagebrush/grasslands. Both the Harrell and Sparks families were experienced in the Spanish system of open-range livestock production in Texas and California before participating in the transfer of this concept to the sagebrush/grasslands. The Sparks family had a long history of association with New Lands agriculture in the southeastern United States before they reached Texas. Such an association implies participation in livestock production on the woodland ranges of the southeast. The ranching enterprise founded by these men was responsible for technological innovations, especially in water development and livestock breeding and husbandry and, therefore, provides a good example of the nineteenth-century evolution of such processes. Finally, John Sparks was a dominant political figure of the time and, as such, left historical records of his activities.

The ranching empire that Harrell and Sparks developed occupied portions of the three major drainage basins of the Intermountain area: Lahontan, Bonneville, and the Snake River. The history of the development of ranching in the Intermountain area also tends to be segregated into these geographical areas. The Snake River sweeps in a great arc through southern Idaho, forming a huge valley bounded on the north by the mountains of the Idaho batholith and on the south by the mountainous lip of the Great Basin. In southeastern Idaho, the Bear River, a tributary to the Great Salt Lake, encroaches into the Gem State. To the west the Owyhee River and its tributaries extend deep in north-central Nevada, draining the vast Owyhee Desert to the Snake. Between the Owyhee and the Bear there are a series of streams which rise in the highlands at the border of the Great Basin and flow north to the Snake. On its rush to the Columbia, the Snake River flows through deep canyons with nearly sheer walls. On top of these walls is a vast, empty plain, clothed in sagebrush, sand, and raw lava flows.

To the nineteenth-century traveler, Idaho was disguised in various ways and, in places, appeared openly hostile. Lacking knowledge of how to approach the vastness of Idaho, early travelers left descriptions of the land which were seldom favorable.[1] In his volume *Astoria*, Washington Irving gave a description of the Snake River Valley and its formidable appearance: "A dreary desert of sand and gravel extends from the Snake River almost to the Columbia. Here and there is a thin and scanty herbage, insufficient for the pasturage of horse or buffalo. Indeed, these treeless wastes between the Rocky Mountains and the Pacific are even more desolate and barren than the naked, upper prairies on the Atlantic side; they present vast desert tracts that must ever defy cultivation, and interpose dreary and thirsty wilds between the

40

habitations of man, in traversing which the wanderer will often be in danger of perishing."[2]

The Snake River Plains consist of a generally flat area bordered by rugged mountains and extending in a curved course concave to the north across southern Idaho. When viewed in the strong sun light of early morning or in the lengthened purple shadows of evening, the plains appear to be absolutely level, and, from many points of view, of ocean extent. The plain of the Snake River is a dissected plain. In places its surface is mildly uneven and in others exceedingly rough owing to the character of the naked lava of which it is composed. It is also marked by many volcanic cones and by broad uplands from vent flows. The channels cut deeply into the Snake River lava and the elevations that rise above the plain are minor features in comparison to the plain itself.[3]

In trying to reconstruct the pristine environment of the Snake River Plains, the brilliant descriptions of the nineteenth-century geologist I.C. Russell are invaluable. "One must become familiar with the characteristics, however, and learn to judge them by their own standards before their beauties are revealed. To the traveler from humid lands, where every hillside is clothed with verdure and every brook flows from a shadowy vale, they will at first seem repellent deserts, on which a long sojourn would be intolerable. When the sun is high in the cloudless heavens, the plains are gray, russet brown, and faded yellow, but with the rising sun and again near sunset they become not only brilliant and superb in color, but pass through innumerable variations in tone and tint. When the approaching dawn is first perceived, the sun is seemingly a great fire beneath the distant edge of the plain. A curtain is quickly drawn aside, revealing a limitless picture suggestive of the view a mariner sometimes has on approaching a bold coast while the actual shoreline is still below the horizon. The distant mountains, rising range above range and culminating in some far off sun-kissed peak, are the most delicate blue, while all below is dark and shadowy. As the sun mounts higher colors deepen, becoming violet and purple, of a strength and purity never seen where rain is frequent. Purple in all its rich and varied shades is the prevailing color imparted to arid lands when the sun is low in the heavens. As the dawn passes the light becomes stronger and the rich hues fade. The mountains recede and perhaps vanish in the all pervading haze, details become obscure even in the immediate foreground and the eye is pained by the penetrating light. The shadows, if canyon walls are near, are sharply outlined and appear black in contrast with the intense light reflected from the sunbathed surface. The light grayish-green leaves of the ubiquitous sagebrush give color, or perhaps more properly lack of color, to the plains and enhance their monotony. In the glare of the unclouded noontide summer sun the plains are featureless, or perhaps their expression is distorted and rendered grotesque or vague and meaningless by the deceptive mirage. During the cloudless summer the glories of sunset are on the earth, not in the sky. As the sun disappears, a well-defined twilight arch arises in the east, the shadow of the earth on the dust particles in the air. As evening approaches there is a gradual change from glare to shadow. The broad plain becomes a sea of purple on which float the still shimmering mountains. The shadows creep higher and higher, until each serrate crest becomes a line of light, margining rugged slopes on which every line etched through centuries by rills and creeks reveal its history. The mountains seemingly grow in stature and unfold ridges and buttresses separating profound depths. One marvels at the diversity and strength of the sculpturing of what but a few moments before appeared flat, meaningless surfaces. The flatland has details everywhere on its surface. The mountains stand boldly forth as sculptured forms of amethyst and sapphire, every line on their deeply engraved slopes, although leagues distant, clearly visible. The last ray of sun winks below the horizon and darkness descends quite suddenly. The cool, starlit summer nights are wonderfully magnificent, the heavens, without a cloud, are filled from horizon to zenith with stars which burn with a steady planetary light, such as is seen in our eastern humid lands only during clear winter weather."[4]

The Sawtooth Mountains of south-central Idaho express in their name their salient features. Following along the Idaho-Utah border, the mountains extend westward to the Jarbidge upland on the Idaho-Nevada border. West of the Jarbidge upland, the Owyhee Desert sweeps down into Nevada with a series of rugged dissected canyons in the desert plain. The Sawtooth Mountains, because of their geographic position and the environment they supported, played an important part in the development of the area. The perennial streams that drained from the mountains supported habitat for beavers which attracted American and Hudson's Bay Company trappers. Milton Sublette and a party of trappers from the Rocky Mountain Fur Company trapped along Goose Creek in 1832. The mountains provided an upland causeway for emigrants to partially cross the Great Basin avoiding the Great Salt Desert to the south and a portion of the Snake River Plains to the north. The mountain environment provided summer ranges for livestock and the streams provided natural meadows that could be enlarged by irrigation.

These pinnacled mountain tops are composed mainly of quartzite, but numerous granite spires and prominent limestone ridges are also sharply defined. The mountains rise boldly to a height varying from 6,000 to some peaks above 10,000 feet. At a distance the mountains appear as a solid wall, but upon entering their canyons, it is possible to wander among and through the higher peaks by a series of relatively low passes. It was by this series of passes the Humboldt Emigrant Trail wound from Fort Hall in southeastern Idaho to the headwaters of the Humboldt and on to California.

Goose Creek is characteristic of the Sawtooth drainages. In this basin an outcropping of sedimentary rock on Trapper Creek contains numerous leaf fossils. Dated on the basis of its stratigraphic sequence, the fossils have been placed in the Miocene epoch of geologic time, some ten to fifteen million years ago. The Miocene was the beginning of the golden age of large herbivores when great herds of camels, horses, and mammoths grazed in western North America. Great rivers drained the present Great Basin to the ocean and the Sierra Nevada and Cascade mountains had not yet risen to rob the Pacific winds of their life-giving moisture and to cast a rain shadow across the Intermountain area. The Trapper Creek flora is composed of sixty-six species representing a forest in which both conifers and broadleaf hardwoods were well-documented. A similar ecological situation today would be the upper Pacific coastal forest at 3,000 feet elevation with its forty to fifty inches of annual rainfall. Some of the plants that compose the present sagebrush/-grasslands evolved from those represented in the Miocene forest.[5]

The mixed coniferous forest represented by the Trapper Creek flora persisted under gradual drying conditions until the close of the Tertiary period. During the Pleistocene or first epoch of our current Quaternary period, the rise of the Sierra Nevada and Cascade mountains brought extreme aridity to the Intermountain area with highly seasonal winter precipitation. This great change brought about the environmental conditions in which the sagebrush/grasslands ecosystem evolved. The Pleistocene epoch or Ice Age is very recent in geologic history and, therefore, the vegetation formation of the sagebrush/grasslands has evolved only recently and may never have reached equilibrium with the newly arid environment before domestic livestock were introduced.

Despite the forbidding appearance of the Snake River Plains, the country was claimed. In 1837, endeavoring to keep the American interest away, the Hudson's Bay Company purchased Fort Hall from New England merchant, Nathaniel Wyeth, and kept it as an outpost on the eastern perimeter of its Pacific division. Evidence of great pasture potential was scattered in every direction around Fort Hall and the settlers had only to notice the opportunity.[6]

Southeast of Fort Hall in 1843, John C. Fremont camped on the southeast Bear River and wrote in his journal: "I can say of it, in general terms, that the bottoms of this river (Bear), and some of the creeks which I saw, now form a natural resting and recruiting station for travelers now and in all the time to come. The bottoms are extensive, water excellent, timber sufficient, the soil good and well adapted to grains and grasses suited to such an elevated region. A military post and civilized settlement would be a great value here, grass and salt so much abound. The lake will furnish exhaustless supplies of salt. All the mountains here are covered with a valuable nutritious grass called bunchgrass, from the form in which it grows it has second growth in the fall. The beasts of the Indians were fat upon it; our own found it a good substance, and its quantity will sustain any amount of cattle, and make this truly a bucolic region."[7]

At Fort Hall the Oregon Trail pioneers had a choice of taking an alternate way to happiness. If Oregon no longer had its original fascination, they could trade that dream for a new one in California. After 1848 the attraction of California and gold became so great the Hudspeth's Cutoff eliminated the need to go to Fort Hall by breaking from the Oregon Trail near Soda Springs, Idaho. Diaries written by the pioneers gave brief descriptions of the area, "Passed over a valley covered with wild wheat as high as my shoulder. It was headed out and looked like a cultivated wheat field."[8] These reports indicated high country for summer range with water available and rich bottomlands with tall grass for winter grazing. In 1855 colonization moved into what was to become the Idaho Territory when a party of Mormons left Utah to establish a mission in the Lemhi Valley. Indian hostility forced the colonists to return to Utah. The Hudson's Bay Company abandoned Fort Hall in the mid-1850s and the Snake River Plains remained in the hands of the Indians with occasional travelers hurrying along the Oregon Trail.

E.D. Pierce found rich gold deposits in August, 1860, on the Clearwater River in the Idaho Territory. By 1861 hundreds of miners were prospecting in the mountains. Several strikes were made extending to within seventy miles of Fort Boise in 1862. Suddenly there was a reason for people to come to Idaho and a market for beef.[9]

In the spring of 1864 William Bryon was engaged in the business of butchering beef and feeding miners in the Boise Basin. He had imported 500 beefs from the Washington Territory. With the onset of winter he drove the animals south toward Nevada, hoping to find a mild valley with tall grass in which he could winter the cattle. He planned to cross the Snake River on the ice, but when he reached the river, it was not frozen. Bryon and his herders suffered from the cold and wind as did the cattle. He was worried because he was stuck against the river in a desert environment below the sagebrush zone with no feed available. The next morning he awoke to find the cattle "full as ticks," but was at a loss to explain what they had eaten. Watching the animals forage, he found they eagerly consumed the herbage of winterfat (*Ceratoides lanata*), a native shrub.[10] Undoubtedly, the value of winterfat has been rediscovered many times, but its ultimate importance in the development of the cattle industry cannot be overestimated.

Joseph Pattee was an agent for Hudson's Bay Company at Fort Hall. He accumulated considerable cattle by trading rested animals for sorefooted animals of emigrants on the Oregon Trail. After the company withdrew from Idaho, Pattee continued to accumulate cattle. In 1867 he decided to push westward to the Raft River Valley and establish a new ranching area. His cattle wintered well in the valley.[11] This operation did not persist, but it showed that cattle could be wintered in the Raft River Valley.

In 1865 near the point where Rock Creek leaves its canyon and ventures out among the rolling hills of the Snake River Plains, James Bascomb established a ranch and stage station. This extended the ranching frontier halfway across the Snake River Plains from Fort Hall. As with the first step into the Raft River Valley, the Rock Creek Station was not destined to last. Bascomb was killed by Chief Buffalo Horn during the Bannock War. The Bannocks were probably the most warlike of the southern Idaho Indian tribes. The tribe was herded onto a reservation at Fort Hall in 1869. This greatly encouraged the spread of ranching. In 1878 a small band of Bannocks stole forty head of beef from George Shoup. He was pasturing the cattle in Lemhi County in southeastern Idaho and killing the animals for jerked beef. The Bannocks moved off the reservation and went on a wide sweep across southern Idaho and eastern Oregon. Their actions became more hostile as they proceeded west, wantonly destroying cattle and killing ranchers.[12] The killing and destruction of stock hardened the attitude of settlers toward the Indians. Living with conflict, fear, and resentment, a great many early settlers developed a harsh rationale about Indians. Typical of this rationale which was passed from generation to generation was the attitude of the pioneer Idaho cattleman, David Shirk: "I want to say right here, and in all sincerity and with undue prejudice, that the only 'good Indians' I ever knew were dead ones. As a race, they are absolutely devoid of all feelings of gratitude, or any kindred spirit. There may be, and probably are, exceptions but as a race they are cruel, heartless, and treacherous as a coyote. Humanitarians may shed their tears over the 'poor' persecuted Red Man, but mine are reserved for their victims, hundreds of whom are heartlessly butchered by these painted and befeathered devils."[13]

If the early ranchers of Idaho had been restricted to sorefooted cattle and occasional drives from Oregon or Washington territories, the vast ranges would have required years to be fully stocked. Instead, the ranges were fully stocked within a decade by importing thousands of cattle from Texas and California to the sagebrush/grasslands. There was a demand for beef in southwestern Idaho in the Silver City mining district. Pioneer cattleman, Con Shea, brought in Texas cattle in 1868. He was on his way to Texas to buy cattle and met Miller and Walters in the vicinity of the Raft River and bought a trail herd they had brought from Texas. Miller and Walters returned to Texas and the next season trailed a herd of stock to the Bruneau River Valley. In 1869, J.G. Shirley and C.S. Gamble reached the Fort Hall bottoms with 3,000 head of Texas cattle. In the mid-1860s Shirley had been one of the ranchers on the Fort Hall bottomlands when the U. S. Government bought the land for Indian reservations. In exchange Shirley received six sections of land on the Raft River at the mouth of Cassia Creek. Later in the fall, attracted by the abundant forage, they moved the herd to the Raft River Valley. The next year Shirley and his associates trailed 10,000 more cattle into the Raft River area. The following year, A.D. Norton and M.G. Robinson proved more venturesome and moved further west to the Rock Creek area near the present town of Hansen, Idaho.[14]

The bottomlands along the Raft River, Goose Creek, and Rock Creek were noted for their dense stands of tall native grasses. On the benches above the creeks were extensive stands of winterfat. There were occasional tracts called "winter parks" where winterfat was abundant and natural hot springs occurred nearby with patches of willows for shelter.[15]

The early settlers noted that sudden winter snowstorms were usually quickly succeeded by thaws during the hours of sunshine. Frequently warm winds, termed "chinook winds," occurred during winter. These winds caused snow, if present, to disappear as if by magic, leaving the plains and lower slopes of the bordering mountains bare.[16]

The first ranches were extremely crude. At most, a pole corral was constructed to hold working horses. The ranchers lived in crude sidehill dugouts. Dimension lumber was not easily available nor would the expense of transportation be justified; therefore, houses were made with whatever materials could be found locally. George Miller spent the winter of 1867–68 ranging his cattle on the Snake River and hired David Shirk to help with the herding. The two men found a place which had an abundance of feed and proximity to the river. They dug down into the riverbank about four feet, deep enough for a room, which when framed in, had proportions of ten by fourteen feet. A large willow ridgepole used for roof support was then covered with smaller brush and ryegrass. Dirt, six inches deep, completed the roof insulation and provided weather protection. A chimney was constructed from rock and dirt and a blanket served as the only door. The house seemed adequate, for seldom did anything inside freeze. Chinook winds often gave unexpected surprises to the dugout dwellers. One winter evening when Shirk and Miller went to bed there was over a foot of snow on the ground. The wind began to blow. Toward morning Shirk awakened, hearing something unusual. Jumping from the bed, he intended to look outside the door but instead found himself standing in four inches of water. The sound that awakened him had been water running down the hill and into the dugout. For the remainder of the winter they did not have to wet the dirt floor to keep the dust down. Outside, the ground that eight hours before had had snow on it was now bare.[17]

At the same time that the first ranchers were struggling to become established in Idaho, a parallel development was taking place in Nevada. The honor of being the first man to winter cattle in Nevada apparently goes to Captain H.A. Parker, a wagon master employed by Ben Holliday. In 1851 Parker wintered in Carson Valley with a train of freight animals. With the draft animals were three milch cows.[18]

Development of the Comstock Lode in 1861 created a demand for beef. The first source of supply was California. The California ranchers were quick to see the advantage of ranching in western Nevada once a local market developed. In 1864 a disastrous drought in California brought thousands of cattle to western Nevada. Rich mineral strikes at Austin in 1864, Pahrangat in 1865, White Pine in 1866, and Eureka and Pioche in 1866 brought increased markets and spread stock growing across Nevada.

Up until 1858 all the cattle in Nevada belonged to the so-called American breeds. These were the common cattle brought by the emigrants from the eastern United States. In 1858 two droves of Spanish longhorns arrived from California. Dorsey and Nottinger delivered 1,500 head to the Truckee meadows and L.B. Drexol brought 2,000 head to Carson Valley. These California longhorns were the offspring of Spanish cattle that had been brought from Mexico to the Spanish missions in Alta, California.

Occupation of the northern two-thirds of Nevada by ranching interests between 1860 and 1880 must rank as the most significant event in the agricultural development of the state. Ranching, following the eastward movement of the mining frontier after the Comstock discovery, usually found its markets in the camps established after new strikes.

The building of the Central Pacific Railroad up the Humboldt River and across northwestern Utah to join with the Union Pacific changed the potential markets of the livestock industry in Nevada. No longer were Nevada ranchers dependent on local mining camps for a market. The local markets required a few animals every week while the completion of the railroad opened the way for the seasonal marketing of

48

large numbers of animals to feed the growing population of California or ship to the midwestern livestock markets.

The Central Pacific Railroad continued from Wells, Nevada, to Tecoma, north of Pilot Peak on the Nevada-Utah border and then across the Great Salt Lake Desert. The railroad route placed Thousand Springs Creek Valley within easy reach of a shipping point and also provided a transportation center for cattle driven from south-central Idaho.

The Humboldt River Valley, from the Big Meadows at its terminal sink to Elko, became the site of the base properties from which ranchers conducted extensive range operations. The Little Humboldt River draining the east slope of the Santa Rosa Mountains and flowing down through Paradise Valley to join the main river at Winnemucca supported another center of livestock production. The Reese River joined the Humboldt at Battle Mountain. The Reese River Valley extended into central Nevada for 150 miles from the Humboldt. On the upper reaches of the Reese the high mountains of central Nevada provided an environment conducive to livestock production. At Carlin, Nevada, ranchers spread south into Pine Valley and north along Maggie and Susie creeks. Near Elko the Humboldt River divides into three major forks. The South Fork, as its name implies, swings south to drain the southwestern face of the Ruby Mountains; likewise, the North Fork swings north to parallel the Independence Mountains; and, the Main Fork or Marys River continues northeast to head against the Jarbidge Mountains opposite the headwaters of the East Fork of the Bruneau River in Idaho. All of the forks of the Humboldt head toward mountain ranges with peaks at least 10,000 feet. Each of the valleys associated with the upper forks of the Humboldt proved to be exceptionally adapted for the culture of cattle. In combination they came to form Elko County, one of the most important range livestock production centers in the western United States.

The Humboldt River route of the California Trail has been called the "high road to the West" and rightly so. The trail left Soda Springs in southeastern Idaho and wound its way across southern Idaho following or twisting through the mountain causeway past the City of Rocks and down Birch Creek and into Goose Creek. The trail led up Goose Creek to the far northeastern corner of Nevada and then down Rock Creek to

Thousand Springs Creek which it followed to its headwaters. Thousand Springs Creek drains into the Bonneville Lake Basin. The emigrants crossed the divide at the head of Thousand Springs Creek to enter the Lahontan Basin and followed down Bishop Creek to Humboldt Wells.

Upper Goose Creek and Thousand Springs Creek are some of the most isolated areas in the modern Great Basin, but in the 1850s they were on the mainstream of the western migration. Many people had the opportunity to see the potential of the environment to support domestic livestock but most saw only the endless sagebrush, dust, heat, and felt only fear of harassment from hostile Indians.

Thousand Springs Creek is similar to hundreds of other drainages in the Great Basin. It starts and ends in obscurity. The headwaters of the drainage are on Antelope Peak and Burnt Crown Mountain north of Wells, Nevada. At first the stream heads north as if striving for the headwaters of Salmon Falls Creek and eventual freedom to the Pacific Ocean. However, this was not to be, and the creek turns south to Bill Downing's H-D Ranch, where Toano Draw enters from the south. Once Thousand Springs Creek passes the H-D Ranch, it swings far to the north, avoiding Tony Mountain and receives the waters of Rock Springs Creek and Crittendon Creek. The stream swings back south to finally emerge from the Toano Range into the valley at Montello. The total length of the creek is about fifty miles with an average drop of eighteen feet per mile. The name Thousand Springs was an obvious choice considering the many springs that occur along the creek. The name was first applied on maps prepared by Lieutenant Beckwith for explorations and surveys for railroad routes. The valley at Montello has several significant features that contribute to livestock production. The oblong valley, some fifty square miles in area, is dominated on the southeast by the austere Pilot Range and the massive 10,000 plus foot Pilot Peak. At the northeastern end of the valley there is a gap between the north end of the Pilot Range and the south end of the Tecoma Mountains. Through this gap extended an arm of pluvial Lake Bonneville, the Ice Age lake of which the Great Salt Lake is a remnant. On the vast alluvial fans that spread from the Pilot Range across the valley at Montello and through the northeastern gap to the Bonneville Salt Flats were thousands of acres of winterfat and other salt desert shrubs suitable for wintering cattle. A second feature of the area that enhanced livestock production was the several thousand acres of saline/alkaline plant communities that occurred where Thousand Springs Creek spilled onto the valley floor. These communities ranged from wet meadows and alkali bullrush (*Scirpus robustus*) marshes to extensive areas of the tall grass Great Basin wildrye (*Elymus cinereus*). This semiwetlands area provided an extremely valuable grazing resource, much more productive than the sagebrush/grasslands. The last feature of the valley that enhanced livestock production was man-made, the Central Pacific Railroad.

Rock Creek and Crittendon Creek extend north from Thousand Springs Creek in broad basins that provide hundreds of thousands of acres of sagebrush rangeland. The Toano Basin extends south from Thousand Springs into the northern end of the Pequop Range and additional thousands of acres of rangeland, including extensive areas of pinyon/juniper woodlands. The town of Toano was an important freighting station before the Central Pacific Railroad was completed. Freight wagons left Toano for the south for the McGill ranches and the eastern White Pine mines. To the north it was six days fast freight to Boise City.

The first "ranch" between Salt Lake City and Carson Valley, Nevada was located near Humboldt Wells and operated under dubious circumstances. Peter Haws, an early Canadian convert of Joseph Smith who fell out of favor with the Mormon settlers in the Salt Lake Valley, established himself on the Humboldt in 1854. Haws's daughter married Carlos Murray, who settled in Thousand Springs Valley. The combined families raised a garden, sold the produce to emigrants, and traded for sorefooted cattle. Dark rumors began that the Haws families aided the Indians in stealing back the animals they traded to the emigrants. Eventually Haws was forced to flee for his life and Carlos Murray and his wife were killed in Thousand Springs Valley by the Indians they were accused of aiding.[19]

Probably the next rancher to try his luck in Thousand Springs Creek Valley was Bill Downing. Bill became disillusioned with the emigrant trail to California and dropped out to rest at a likely-looking spring in the upper part of Thousand Springs Valley. The surroundings grew on him and he decided to stay. He combined trading for sorefooted cattle with tending a garden and selling supplies he bought in Wells. He branded his stock with the H-D brand and called his ranch the Ox-Yoke Ranch.[20]

In the late 1860s Bill Downing started to have neighbors, or at least someone living within a three-days horseback ride. Colonel J.B. Moore had commanded Camp Ruby in Ruby Valley on the east side of the Ruby Mountains during the Civil War. He used the troops under his command to help develop a ranch site. After the War he was quick to purchase Texas longhorns to stock his ready-made ranch. Colonel E.P. Hardesty had fought on the other side during the Civil War, but he also saw the opportunity for ranching in Elko County and brought cattle from Texas to near Bishop Creek north of Wells.[21]

Large and small ranchers had found a fortunate combination of near-free grazing, easily monopolized water sources, growing markets in California and the Midwest, and a stable, easily controlled political climate sensitive to the requirements of an increasingly ranching-oriented economy.[22] In 1869 the editor of the *Elko Independent* wrote, "A new brand of business has been gradually growing up in raising and exportation of beef cattle—fattening of cattle and driving them to California has become an important business—large herds can be seen all along the valley of the Humboldt and its tributaries—In time, stock raising will be in sound value and extent greater than that of the production of precious metals."[23]

The time was right for the large-scale exploitation of the environment for livestock production. What was needed was an entrepreneur with the skill to manage livestock, the capital to take advantage of the situation, and the guts to take the risk. Such a man was Jasper Harrell. Born near Augusta, Georgia, on August 16, 1830, Jasper grew up on the family plantation under relatively wealthy conditions. The Harrells were a family of slave owners engaged in cotton culture under plantation agriculture. Jasper joined the rush to California in 1850 by ship to Panama and then, again by ship, on to San Francisco. He mined at various locations in California and finally settled in Tulare County in 1856. Jasper developed extensive ranching interest, headquartered on Cross Creek near Visalia. Besides ranging cattle on the California annual range, Jasper Harrell brought thousands of acres under cultivation to cereal grain. Mr. Harrell reinvested his money in real estate in the San Joaquin Valley in California. Jasper Harrell married Martha Bacon in 1857 and in 1861, a son, Andrew Jasper, was born. A.J. Harrell was later to play an important part in his father's ranching operations.[24]

Jasper Harrell was nicknamed "Barley" Harrell supposedly because

52

of the sack of barley he always carried behind his saddle to provide grain for his horse.[31] His land was producing thousands of bushels of barley on ranches in the San Joaquin Valley so he had a ready supply. "Barley" Harrell was very much a frontier figure. He rode the ranges of northeastern Nevada and south-central Idaho with a lever-action rifle in a saddle scabbard and six-shooter in his holster. He paid his range crews every three or four months, carrying the gold and silver coins in his saddlebags. He timed his arrival at the cow camps for dinner time. After dinner Jasper would gamble with the crew and usually win back a large portion of the money he had paid in wages.

Jasper Harrell probably bought his first ranch in Nevada in 1870. He bought cattle in Texas in 1870 and had them delivered to Thousand Springs Creek Valley. Among the cowboys employed to drive the herd was Louis Harrell, a nephew from Georgia who decided to leave the post-Civil War South and take his chances with his uncle in the Far West.[25] There was a ranch located at or near Tecoma belonging to a Mr. Armstrong which Jasper Harrell absorbed in a partnership.

Jasper Harrell was well established on Thousand Springs Creek with extensive winter and spring/fall range available. The ranching operation lacked summer range. To have summer range the landscape had to have high-elevation mountains. The landforms close to Thousand Springs Creek lacked the necessary elevation. The topography and outline of the landforms of Thousand Springs Creek Basin are the product of tertiary volcanic action influencing extremely old Paleozoic and pre-Cambrian sedimentary rocks. The mountains rise to 8,000 feet, some 3,000 feet above the valley floor. The ranchers established on the upper Humboldt were exploiting the Ruby Mountains as those on the north fork were using the Independence Mountains. Jasper Harrell sent his foreman, James E. Bower, north to upper Goose Creek in 1872 with a herd of 3,000 cows to search for additional summer range. One day on a short exploring trip in the vicinity of Goat Springs, Bower was able to see far to the north from a high ridge and obtained his first glimpse of the Snake River Valley. Bower could not resist riding down from the mountains to assess his new discovery. He rode down through pristine communities of bluebunch wheatgrass and Idaho fescue. The aspect was that of a grassland not a shrub steppe. Upon reaching Cottonwood Creek, he met two Rock Creek cattlemen, A.D. Norton and M.G. Robinson. These two cattlemen were probably feeling lonesome and very vulnerable to hostile Indians at their remote ranch. They were quick to point out the potential of the country for stock growing. Most importantly, Norton told Bower of the wintering potential of the lower portions of the Snake River Plains. If Norton and Robinson had realized the growth potential of the cattle operation that Bower represented, they may have been more subdued in singing the virtues of the south-central Idaho ranges. J.E. Bower hurried back to Nevada and told Jasper Harrell of the ranges he had visited. Harrell immediately started buying Texas cattle and pushing them north to establish a claim on the vacant range.[26]

Jasper Harrell became established on Goose, Rock, and Salmon Falls creeks, on both sides of the Nevada-Idaho boundary. One of his favorite ranches became the so-called Winecup field on Goose Creek. It is easy to visualize the charm of the Winecup field, a clear, fast-flowing perennial stream, meandering through emerald-green meadows that contrast with the snow-white volcanic ash bluffs that abruptly edge the valley. The foothills of the Goose Creek Basin are partially clothed with pinyon/juniper woodlands. The round mass of Mahogany Butte rises to the northwest. The higher mountains support patches of lodgepole pine (*Pinus contorta*) and subalpine fir (*Abies lasiocarpa*).

The Winecup field is named for the ⋃ brand. There is no question that Jasper Harrell used this brand, but its origin is in doubt, being variously given as Wyoming or Texas. There is also the possibility he brought it from California. Harrell also used the Shoesole brand ⊂⊃, which was the fourteenth brand recorded in Elko County, registered on May 1, 1873. The Shoesole supposedly originated in Wyoming where it was known as the Indian Moccasin. Jasper Harrell's horses were branded with a smaller version of the Winecup brand on the left shoulder. The earmark of the outfit was right ear cropped on top and left ear cropped under.[27]

In 1871 there occurred an extremely dry winter in California and thousands of cattle were driven to Nevada. California was changing from an open-range area to dry land cropping of cereal grains during the early 1870s. In 1873 a fence law was enacted in that state making owners of livestock liable for damages caused by their animals. This spelled the end of the large, open-range operations in the central valleys of California and many large ranchers moved to the Great Basin.[28]

The capital requirement for the type of expansion that Jasper

Harrell was engaged in during the 1870s was minimal. J.O. Oliphant estimated for the Pacific Northwest the capital requirement for engaging in large-scale livestock production outside of the cattle was $1.12 per brood cow based on units of 5,000 head. Jasper Harrell pumped the profits of almost two decades of successful ranching, grain farming, real estate investments, and banking in California into his Great Basin livestock enterprises.[29]

The range cattleman of the 1870s ran a frontier operation with very little capital. The open range was a practical proposition for he could not buy the land necessary for the great herds to graze even if he wanted to. Most ranchers preferred to move on if the range became depleted. They were prepared to keep ahead of settlement, unhindered by ownership of depleted ranges.[30] The failure of the great banking house of Jay Cook and Company precipitated a national depression in 1873. Jasper Harrell had the advantage of a rapidly expanding market to pull his range cattle operations through this depression. The silver mines of the Great Basin were booming and the population of California was increasing rapidly. Because of its geographical position, history of settlement, and physical environment, the Great Basin was the last natural grazing land to be exploited in western North America and, most probably, the world. The exploitation pageant now had its chief characters, and the saga continued across the sagebrush/grasslands.

Notes

1. Byron D. Lusk, "Golden Cattle Kingdoms of Idaho," M.S. thesis, Utah State University, Logan, Utah, 1978.

2. Washington Irving, *Astoria*, (New York: G.P. Putnam, 1849).

3. I.C. Russell, *Geology and Water Resources of the Snake River Plains of Idaho*, (Washington, D.C.: GPO, U. S. Geological Survey, 1902).

4. In order to encompass a single day, the description is a composite of I.C. Russell's description.

5. D.I. Axelrod, "The Miocene Trapper Creek Flora of Southern Idaho," *Geol. Science*, Vol. 51, 1964.

6. Lusk, "Golden Cattle Kingdoms."

7. J.C. Fremont, *Report of the Exploring Expedition to the Rocky Mountains*, microfilm edition, University Microfilm, Inc., Ann Arbor, Michigan, 1977.

8. H.R. Cramer, *Hudspeth's Cutoff*, (Burley, Idaho: Private Printing, 1969).

9. Lusk, "Golden Cattle Kingdoms."

10. *Idaho Sunday Statesman*, 2 February 1941, Boise, Idaho.

11. A.C. Anderson, ed., *Trails of Early Idaho: The Pioneer Life of George Goodhart*, (Caldwell, Idaho: The Caxton Printers, 1940).

12. D.O. Hyde, *The Last Freeman*, (New York: The Dial Press, 1973).

13. M.F. Schmitt, ed., *The Cattle Drives of David Shirk*, (Portland, Oregon: Champoeg Press, 1956).

14. Adelaide Hawes, *The Valley of Tall Grass*, (Bruneau, Idaho: Private Printing, 1950).

15. "Mimeograph History of Minidoka National Forest," (Twin Falls, Idaho: Forest Service, U. S. Dept. of Agric.). Also see Alexander Toponce, *Reminiscences of Alexander Toponce*, (Norman, Oklahoma: University of Oklahoma Press, 1971).

16. C.W. Gordon, "Report on Cattle, Sheep and Swine," *Supplementary to Enumeration of Livestock on Farms in 1880*, 10th Census of the United States, Vol. III, (Washington, D.C.: GPO, 1880).

17. Schmitt, "Cattle Drives of David Shirk."

18. Gordon, "Report on Cattle, Sheep and Swine."

19. D.L. Morgan, *The Humboldt Highroad of the West*, (New York: Farrar & Rinehart, Inc., 1943).

20. Information on Bill Downing is from the narrative of Captain C.H. Roland, Wells, Nevada, on file with Nevada Historical Society. The H-D Ranch was located at the site of the present-day Winecup Ranch.

21. E.B. Patterson, L.A. Ulph, and Victor Goodwin, *Nevada's Northeast Frontier*, (Sparks, Nevada: Western Printing and Publishing, 1969). See also, V.S. Truett, *On the Hoof and Horn in Nevada*, (Los Angeles, California: Gehrett-Truett-Hall, 1950).

22. J.M. Townley, "Reclamation in Nevada, 1850–1904," Ph.D. diss., University of Nevada, Reno, Nevada, 1976.

23. *Elko Independent*, 15 December 1864, Elko, Nevada.

24. J.M. Guinn, *History of the State of California and Biographical Record of the San Joaquin Valley*, (Chicago, Illinois: The Chapman Publishing Company, 1905).

25. Personal communication to James A. Young from N.J. Harrell, Louis Harrell's son, Twin Falls, Idaho, July 1979.

26. C.S. Walgomott, *Reminiscences of Early Days*, Two Volumes, (Twin Falls, Idaho: Private Printing, 1927).

27. N.L. Bowman, *Only the Mountains Remain*, (Caldwell, Idaho: The Caxton Printers, Private Printing, 1955).

28. Townley, "Reclamation in Nevada."

29. J.O. Oliphant, *On the Cattle Ranges of the Oregon Country*, (Seattle, Washington: University of Washington Press, 1968).

30. C.M. Owens, "Early Cattle Raising in Wyoming," M.A. thesis, Colorado State Teachers College, Greeley, Colorado, 1932.

1. The Great Basin is a land
   of contrast. Early settlers
   saw rushing rivers with
   juniper woodlands.

2. They also saw stark, salt desert playas.

3. The Lahontan sand dunes formed in the salt desert.

4. The ranchers shared the land with many animals, among them the coyote shown here in tall crested wheatgrass.

5. Great Basin sagebrush/grasslands occupy alluvial soils in the valleys and at the bases of the mountains. Pinyon/juniper woodlands with mountain brush communities form still farther up the slope.

6. Bluebunch wheatgrass, an important forage grass in the sagebrush/grassland, grows at the edge of a juniper woodland.

7. Extensive stands of Indian ricegrass occur in the salt desert and are used to winter cattle.

8. Winterfat also has been used to winter cattle. It grows most abundantly in the transition zone between salt desert and sagebrush communities.

9. Greasewood shrubs grow on dunes in the Carson Desert.

10. Polygonal cracks appear on the surface of drying playa. I. C. Russell compared their color and texture to fine Italian marble.

11. The sagebrush/grasslands contain many perennial grasses including needle-and-thread grass, which is often found on sandy soils. The light-colored shrub is horsebrush.

12. This site in high range condition shows the mixture of shrubs, forbs, and grasses.

CHAPTER
3

LEFT-HAND TRAIL TO HELL

The cowboys from Texas who delivered cattle to Nevada must have thought they were riding down the left-hand trail to hell when they dropped the herds of Longhorns into the arid basins left by Lakes Lahontan and Bonneville. It would be difficult to find a more stark and forbidding landscape in western North America.

Describing his travels in the Lahontan Basin in the 1880s, I.C. Russell noted, "The valleys or plains separating the mountain ranges, far from being shady vales, with life-giving streams, are often absolute deserts, totally destitute of water, and treeless for many days' journey . . . . Many of them have playas in their lowest depressions—simple mud plains left by the evaporation of former lakes—that are sometimes of vast extent. In the desert bordering the Great Salt Lake on the west and in the Black Rock Desert of northern Nevada are tracts hundreds of square miles in area showing scarcely a trace of vegetation. In winter, portions of these areas are occupied by shallow lakes, but during the summer months, they become so baked and hardened as scarcely to receive an impression from a horse's hoof, and so sun-cracked as to resemble tessellated pavements of cream-colored marble. Other portions of the valleys become encrusted to the depth of several inches with alkaline salts which rise to the surface as an effervescence and give the appearance of drifting snow."[1]

In a humid environment, forage production depends on periodic rainfall during the growing season. As ranching proceeded across North America in the nineteenth century, ranchers found increasingly arid conditions. In the Intermountain area, it rained only infrequently during the growing season. Would-be ranchers had to adapt to the physical and biological restraints of the sagebrush/grasslands environment. Ranch headquarters located in the sagebrush/grasslands had environmental constraints. Permanent water, meadows, irrigable land, and the proximity of winter and summer ranges affected where ranches were located and what land was purchased.

Despite their abundance, the sagebrush/grasslands had disadvantages as environments for production of red meat from cattle. The

59

green feed period is short—spring and early summer. As calves born in the spring become independent grazers in midsummer, the native grasses become mature, and quality rapidly declines. To avoid this loss in forage quality, livestock would follow the green feed up the mountains and graze on high mountain summer ranges.

Winters are much too severe to allow year-long grazing. As a result, transhumance (man and livestock moving with the seasons) agriculture developed in the Intermountain area; sagebrush ranges were used in the spring and fall, mountain ranges in the summer, and salt desert ranges in the winter. Cattle were wintered on the margins of the salt deserts and winter grazing was looked upon as very profitable.

After the Ice Age, Lake Bonneville was 19,750 square miles in area, and had a maximum depth of 1,000 feet. The Great Salt Lake is a remnant of this lake. Lake Lahontan covered 8,422 square miles and the deepest part, the present site of Pyramid Lake, was 866 feet deep.[2] The accumulation of soluble salts left in these old lake basins earned them the name of salt deserts. Not all of the soils in these basins are salty, but all the basins receive very little rain. Often four inches or less annual rainfall occur in the centers of the basins. Precipitation occurs during the cold winter months when temperatures prohibit plant growth.

Despite their apparent simplicity, playas play an important role in the ecology of Intermountain environments. Many of the playas are deflated (i.e., have their sediments blown away) during the dry portion of the year. Even during winter months, it does not take long for the surface of the playas to dry. In the winter, snow cover melts rapidly on the playas because of the surface's salt content. The clay-textured soil particles on the surface of the playas flow together when wet, making it impossible for moisture to percolate into the soil. If a considerable amount of moisture accumulates or flows onto the playa surface, shallow lakes form. Smaller amounts of moisture evaporate from the surface of the playa and leave crystals of soluble salts on the surface soil. The wet playa glistens, then turns dirty gray as the film of water evaporates. When completely dry, under a bright sun, the playa's salt crystals are dazzling to the point that it is impossible to look across the flats without eyes watering.

The crystal dazzle does not last very long, for soon the winds start to stir across the playas. The white playas are like huge mirrors reflecting the sun's energy back into the atmosphere. The reflected rays merge to mirror the sky and create shimmering mirages which offer false hope of water where there is none. The mirage cuts off the base of the towering mountain. From far out on the playa, the horizon is composed of spectral mountains that appear to float in space and continually retreat.

The reradiation of energy heats the atmosphere and soon whirl-winds start to spiral across the playa surface. These swaying and bending columns, often two or three thousand feet high, rise from the plains like pillars of smoke, and form characteristic features of the desert. The energy in these columns is surprising. More than one cowboy has chased his hat across the salt flats.

At the end of a day, the north winds blow across playas which have changed greatly in appearance. The white dazzle of crystals is gone and in its place is a dull cream color. The salts have been redistributed in selective patterns. Geologists who first studied the silver ore deposits of the Great Basin were impressed by the occurrence of silver chloride (or horn silver) on the surface. They attributed this horn silver to salt (NaCL) being recycled by winds off the playas.[3]

Besides salt, large quantities of soil also move off the playas. Mud dunes are formed when salts crystallize on the playas after rain or snow and cement soil particles together into structures the size of salt grains. These salt-soil structures obey the aerodynamic rules of sand grains. They bump and bounce across the playa surface in a wind-driven process called saltation. Once these large particles are driven off the playas and reach areas where there is some vegetation, they are trapped, forming dunes which look like sand dunes. When it rains on the mud dunes, the salts dissolve, the clay particles are freed, and the dunes become mud mounds which make travel virtually impossible.

The vegetation of these vast old lake bed environments is domi-nated by low, widely spaced shrubs. Most of these shrubs belong to the Goosefoot family of Chenopodiaceae. One of the most valuable shrubs for wintering livestock in the salt deserts is winterfat. This white wooly shrub has always suffered from a number of synonyms. It is widely known as white sage. This is unfortunate, because of confusion with *Artemisia cana* which is also known as white or silver sagebrush.

The Idaho butcher who accidentally discovered the grazing value of winterfat started a trend. A shrub that cattle readily grazed by preference and that supplied an adequate diet was rare, and after the first hard frost in the autumn, cattle appeared to prefer winterfat herbage. Livestock eagerly eat its prolifically produced and highly digestible seeds. However, winterfat is not tolerant to excessive grazing.[4]

Besides winterfat, other members of the Goosefoot family, especially the saltbush genus, characterize the salt desert environment. The saltbush genus embraces about 150 species. About sixty of these species occur in the western United States. The Great Basin, with thirty-two species, appears to be the center of distribution for the western species. New habitats opened up in western North America after the recession of Lakes Bonneville and Lahontan and allowed a host of new species of saltbush.[5]

Of the valuable browse species, fourwing saltbush is one of the most important. Fourwing is a freely branched shrub, six to ten feet high, with grayish-white stems and leaves and rigid twigs. The name fourwing is derived from the shape of the fruits. The seed crop of fourwing saltbush is eagerly devoured by livestock when the fruits are mature. Fourwing saltbush is found over vast areas of the salt deserts, but it generally is mixed with other species and rarely dominated. An exception is areas of sand dunes where the genetically different Gigas (or giant) form of fourwing saltbush is found. Fourwing saltbush also occurs in patches around salt flats which are subjected to seasonal flooding.[6]

Shadscale occurs as the dominant species over millions of acres of rangeland within the sagebrush/grasslands. In contrast to winterfat or fourwing saltbush, shadscale is not a preferred browse species. After shadscale plants lose their fruits in the fall, the branches of the short inflorescences on which the fruits were borne become rigid and then spinelike. Such naked spiny branches persist for several years, and provide considerable protection for the shrub against browsing animals. The leaves and fruits of shadscale drop off in autumn and are collected in soil surface depressions or form little wind drifts behind the bushes. These piles of leaves and seeds are the first thing livestock eat when turned onto salt desert winter ranges.[7]

Shadscale is the dominant species on many soils that are not exceptionally salty. These sites are dominated by shadscale instead of sagebrush because of lack of moisture. The shadscale plants are naturally spaced at wide intervals in balance with the limited potential of the sites. Once the shrubs establish, they accumulate and deposit soil particles under their branches.[8] The small mounds of soil under the shadscale plants increase the area they have available for root growth.

The soil surfaces between the shadscale plants are stabilized with desert pavement, a network of small rocks left on the surface when fine salt particles are eroded by winds and rain. The rocks are covered with desert varnish and provide the background colors that characterize the landscape. Sometimes the desert pavement is composed of stones. When the pavement stones are derived from black basalt, the interspaces are as dark as burned landscapes, and provide a preview of the popular concept of hell.

The only sagebrush species that is abundant in the salt deserts is budsage. Budsage has a preference for the salt desert environment and the distinction of being a spring bloomer. In February, when the mountain ranges are still firmly locked in winter snows and the nights are bitterly cold in the desert valleys, budsage begins to leaf out with vivid green leaves. Budsage occasionally coexists in winterfat communities as an understory shrub. During most of the year the winterfat communities are the wooly gray-green color of winterfat herbage. In the early spring the startling green of the budsage leaves followed by their clustered yellow flowers makes the desert traveler wonder where in the world this new species came from. At the first hint of warm temperatures, the budsage leaves turn brown and wilt. The leafless branches are virtually invisible until the next spring. The browse of budsage is valuable food for cattle. Its early growth when other forage is dormant and dry is especially attractive to wintering livestock. In the 1890s, budsage was considered one of the most important forage species for wintering livestock in the Red Desert of Wyoming.[9]

No hard and fast rule limited cattle to the salt desert areas during the winter. The cattle were free roaming, and between storms they drifted back into the sagebrush zone to graze and browse on preferred shrubs such as bitterbrush and cliffrose. The sagebrush zone deserts were safe from deep and prolonged snow. Once the cattle were forced onto the salt deserts by winter storms, their movements were restricted by water distribution. There is virtually no flowing surface water in salt deserts, so cattle were restricted to areas near springs.[10]

To understand the wintering of cattle on the salt desert, one should differentiate between free-roaming cattle and cattle forced to use salt desert areas. If cattle were free to drop down into the desert valleys when winter storms struck the sagebrush, access to the salt desert served as a safety valve to prevent excessive winterkills. In contrast, when cattle were driven out onto the salt desert and left at an isolated spring to forage for themselves, the salt desert could become a deathtrap. The salt deserts were an escape from winter precipitation, but not from winter cold. Cattle can withstand a great deal of cold if their rumens are full. If placed on a starvation diet, however, cattle are susceptible to extreme cold and the salt deserts are cold deserts.

Wintering cattle in the salt deserts was somewhat a matter of a square peg in a round hole. The predominant forage species are shrubs, and cattle are essentially grazers of grasses. Most of the valuable browse species, except for winterfat, are spiny species that are protected from wholesale grazing by cattle. Cattle have a high daily water requirement and limited distance of travel from watering points. In contrast, sheep are selective browsers with a much lower daily water requirement. In fact, sheep can graze on salt desert ranges and depend on skiffs of snow for water.[11] This discovery ranks as one of the most important innovations in livestock husbandry in the nineteenth century.

The only forage grass that was abundant on non-meadow areas in the salt deserts was Indian ricegrass (*Oryzopsis hymenoides*). Accounts of nineteeth-century grazing in the salt deserts often refer to this species as sandgrass. On winter range areas, Indian ricegrass is highly preferred by all classes of livestock. It is rated as good to very good forage for cattle. The plants produce an abundance of herbage which cures well on the stalk and is very nutritious. The plump seeds are also high in food value and are sought by grazing animals. Stockmen of the nineteeth century regarded this grass highly as winter forage. They called it a "warm feed" because of its high value for sustaining livestock during severe winter weather.[12]

Water is vital to the use of Indian ricegrass winter ranges. Sheep can go without water for long periods when the feed is succulent. Horses on range frequently water only at three-day intervals. Cattle have to water every day to perform efficiently. When moisture occurs on the winter ranges small puddles of water form on depressions in the playas. The range-wise cows quickly extend their grazing area and work around these puddles until the last drop of foul, alkaline water is consumed or evaporated.

Indian ricegrass was first grazed near the water holes. Gradually the cattle enlarged a circle around the watering points until all the herbage within a four-mile radius was consumed. In the depths of the Carson

Desert this type of winter grazing is still practiced. At distances greater than four miles from water the cows become much more selective grazers of Indian ricegrass. Less and less of the coarser portions of the plants are consumed. At ten miles from water, the most vigorous cows make only quick passes through the Indian ricegrass stands, consuming the seeds that persist on the multi-branched seed stalks.

If moving down the depths of the desert valleys to escape the cold and snow of winter in the sagebrush was akin to dropping into the ranges of hell, then ascending the mountains to summer ranges must have been the cowboy's version of a journey to heaven. I.C. Russell described May Day as an event celebrated during August in the mountains that rim the western edge of the Great Basin.[13] On the western edge of the Intermountain area it was possible to escape the sagebrush environment by moving to the Sierra Nevada, Cascade, Steens, or Warner mountains. Across the northern portion of the Intermountain area, the mountains of northeastern Oregon and the Idaho batholith provided summer range. On the east the abrupt boundary of the Wasatch Front and the outlying ranges of the Rockies provided summer range.

The mountains that formed the headwaters of the Humboldt Ranges in the northeast, the Toiyabe and Shoshone ranges at the head of Reese River to the south, and the Santa Rosa Range at the head of the Little Humboldt have sufficient highland areas to support extensive areas of summer range and even alpine areas. The many lower mountain ranges that subdivide the Great Basin support better-than-average rangelands, depending on elevation and latitude. The mountains that rim the Intermountain area support mixed coniferous forests with many species of evergreen trees. In the Great Basin even the high mountains lack extensive areas of coniferous forests. From the Truckee River in western Nevada to southeastern Idaho and southward through Nevada and Utah, the lower slopes of the mountain ranges have open woodlands of single-leaf pinyon and Utah juniper commonly called pinyon/juniper (or P J) woodlands. Both species are multi-trunked and bushy.

The pinyon/juniper woodlands are not lush, shaded forests. Pinyons with their irregularly shaped crowns and thickened bases hunker down to the harsh, rocky soils. The trees give the appearance of having had to fight for survival in a generally treeless environment. The needles of single-leaf pinyon are sharp and flecked with resin droplets. If an unaware traveler brushes against a pinyon, the immediate reward is a sharp prick and a lasting coat of pitch.

The seeds of pinyon were an important part of the diet of Indians native to the central Great Basin. The seeds were eaten raw, roasted, cooked, and were preserved in the form of pinole, a finely ground flour. Pinyon resin was used as chewing gum as well as for mending, cementing, and waterproofing. In the fall when the juniper berries are ripe, the pinyon/juniper woodlands are alive with the harsh cries of pinyon jays who feed on the berries.

Both pinyon and juniper species, especially their seedlings and young plants, are susceptible to wildfires. During the 1964 firestorm in Elko County an aerial photograph of the Palisade portion of the fire showed that pinyon/juniper woodlands growing in the valleys and north slopes were burned over and killed. Those growing on steep, south slopes were unburned. Those slopes did not have sufficient understory vegetation to carry the fire from tree to tree. All age classes of trees from century-old patriarchs to seedlings were present. The trees that burned were relatively even-aged stands. These trees had invaded the sagebrush/grasslands, moving out from the true pinyon/juniper stands growing on fire-safe sites. This process has apparently been going on for thousands of years since the close of the Pleistocene.[14]

Above the pinyon/juniper woodlands on the interior mountain ranges, extensive areas of mountain brush occur. These areas form summer range for cattle and mule deer. They appear to be a continuation of the sagebrush/grasslands. Although big sagebrush is a major component of these communities, it is usually represented by the subspecies, mountain big sagebrush, which is more highly preferred by browsers. Besides mountain big sagebrush these communities contain the valuable browse species bitterbrush, curlleaf mountain mahogany, and species of snowberry. These mountain brush communities support productive grasslands dominated by bluebunch wheatgrass or needlegrass species and Idaho fescue on the north slopes.

Mountain meadows play an important part of summer range environment. Fertile soils and a good soil-moisture balance from a high water table contribute to a productive potential much greater than on adjacent sagebrush ranges. These meadows range from a few square feet to several hundred acres in area. The meadows are dominated by tall grasses such as Nevada bluegrass and meadow barley. Under the tall grasses is a tuff sod of sedges, wiregrass, and mat muhly. Forbs include Rocky Mountain iris and pale agoseris.[15] These meadows often are called "stringer meadows" because they form narrow strips along water courses through a generally gray, shrub-dominated environment. Meadows are important for sagegrouse, the "sage chicken" that is characteristic of these environments.[16]

Above the mountain brush, on mountain ranges that extend above 9,000 feet in elevation, scattered stands of five-needle pines occur. These are the ancient bristlecone, white bark pine, and limber pine. Aspen groves are also characteristic of summer ranges. Quaking aspen is one of the few deciduous trees widely distributed in the uplands of the sagebrush/grasslands. On major mountain areas extensive areas of aspen groves, both trees and aspen shrubs, provide browse and shelter. In the shade of the aspen patches and near perennial snowbanks grow tall larkspur plants. These violet-blue flowers are deceptively beautiful. Larkspur plants contain poisons which are especially deadly to cattle. The greatest loss of cattle occurs early in the season because larkspurs start growth before other forage species. Sheep are less susceptible to the poison, and are seldom poisoned in the field. Tall larkspur poisoning limits the use of some summer ranges by cattle.[17]

Under pristine conditions the high mountains of the Intermountain area supported the desert bighorns. Three subspecies of mountain sheep are thought to have occurred in Nevada—the desert bighorn, the California bighorn, and the Rocky Mountain bighorn. For feeding routes, bedding grounds, and all-round living quarters, the bighorn prefers the roughest and most precipitous country, on or near the mountaintops. Even when forced to travel in lower country, the sheep gravitate to high spots that allow them to see the surrounding country. Bighorns are browsers and grazers of forbs and grasses. Each band has a regular feeding route which usually terminates at a bedding ground. Snow and winter cold cause the bighorns to move downslope to desert ranges. For drinking, where a choice exists, the sheep select springs or water holes in the most inaccessible, higher parts of the mountains.[18]

The elk (or wapiti), the largest member of the deer family on sagebrush ranges, was almost absent from the Intermountain ranges under pristine conditions. In the Pacific coast region, elk occupied the Coast and Cascade ranges from Vancouver Island southward through

Washington, Oregon, and into southern California. Curiously enough, most of Arizona and Nevada and parts of Utah, Oregon, and Washington appear not to have included elk.

The only historical mention of elk in Nevada comes from Captain J.H. Simpson's journal of his exploration for a wagon route from Camp Floyd, Utah, to Genoa, Nevada. On July 20, 1859, Simpson says, "An elk was seen for the first time yesterday in Stevenson's Canyon, and one today in Red Canyon, also a mountain sheep for the first time." Red Canyon is in the Snake Mountains just north of Wheeler Peak and Stevenson's Canyon is west of the Schell Creek Range, both in eastern White Pine County.[19]

In much of the Great Basin the quantity and quality of summer range often determines population size of mule deer herds. In the more arid portions of the Intermountain area, only the highest mountains provide suitable habitat for mule deer. Thus, the mountains of northeastern Nevada provided excellent mule deer habitat. Many settlers looked on the big game as a source of winter meat. For example, Utah residents from Brigham City made hunting trips to the mountains of south-central Idaho to kill mule deer. In 1883, T.D. Calahan and John Strude returned from the mountains of Owyhee County with forty-five mule deer they had killed.[20]

An almost universal characteristic of summer ranges in the sagebrush/grasslands is that they are steep and rugged. There are very few level uplands that provide summer ranges. Cattle graze level land to their best advantage and tend to congregate on more level areas such as stringer meadows. Cattle can traverse amazingly steep land, but the willingness, not the ability, of an animal to get into the steep areas to graze is what is important. Sheep are better adapted to grazing steep topography. Their smaller size enables them to negotiate steep areas with less difficulty than larger animals. Since they are ordinarily under the control of a herder, they can be forced to graze steep slopes.

The winter and summer escape from the sagebrush/grasslands ranges to the desert or mountains enhances the potential of these ranges for the culture of cattle. Successful ranch management strikes a balance of these resources. For a given amount of sagebrush/grasslands, one must have a balance of winter and summer range. Later the requirement of hay lands was added to this balance.

Satellite winter and summer ranges have some basic weaknesses in the culture of cattle in the cold desert. First, on winter ranges the forage resource is mostly browse species and use of this browse is limited by lack of water. Sheep are more efficient users of browse and require less water than cattle, making sheep better adapted to wintering on the salt deserts. Secondly, the mountain ranges while providing a desirable escape from the summer drought in the sagebrush can also be rugged, steep, and infested with tall larkspur. Sheep are better adapted for grazing on steep, subalpine slopes than cattle and usually are not affected by tall larkspur poisoning. In the 1870s and 1880s these appeared to be small points, but they left the door open for the severe competition between cattle and sheep ranchers that followed.

Notes

1. I.C. Russell, *Geological History of Lake Lahontan*, (Washington, D.C.: GPO, U. S. Geological Survey, 1885).

2. Ibid.

3. S.G. Houghton, *A Trace of Desert Waters*, (Glendale, California: The Arthur H. Clark Company, 1976).

4. W.A. Dayton, *Important Western Browse Plants*, (Washington, D.C.: GPO, U. S. Dept. of Agric., Misc. Bull. No. 101, 1931).

5. *Range Plant Handbook*, (Washington, D.C.: GPO, Forest Service, U. S. Dept. of Agric., 1937).

6. H.C. Stutz, "Explosive Evolution of Perennial Atriplex in Western America," pp. 161–68, in *Intermountain Biogeography: A Symposium*, (Provo, Utah: Great Basin Naturalist's Memoir No. 2, 1980). H.C. Stutz, J.M. Melhy, and G.K. Livingston, "Evolutionary Studies of Atriplex: A Relic Gigas Diploid Population of *Atriplex canescens*," *Amer. J. of Botany*, (1975) 62:236–48.

8. J.G. Smith, *Fodder and Forage Plants Exclusive of the Grasses*, (Washington, D.C.: GPO, Div. of Agrostology, Bull. 2, U. S. Dept. of Agric., 1900).

9. W.D. Billings, "The Shadscale Vegetation Zone of Nevada and Eastern California in Relation to Climate and Soil," *Amer. Midl. Nat.*, (1949) 42:87–109.

10. Avon Nelson, *The Red Desert of Wyoming and its Forage Resources*,

(Washington, D.C.: GPO, Division of Agrostology, Bull. 13, U. S. Dept. of Agric., 1892).

11. L.A. Stoddart, and A.D. Smith, *Range Management*, (New York: McGraw-Hill, 1943). Cows require 7.5 to 12 gallons of water per day and sheep 0.2 to 1.5. Generally, 5 sheep are considered equal to 1 cow in forage requirements, making the maximum water requirement the same as the minimum cow requirement.

12. "Range Plant Handbook."

13. I.C. Russell, *Quaternary History of Mono Valley, California*, (Washington, D.C.: GPO, U. S. Geol. Survey 8th Ann. Rpt., 1886–1887, Pt. 1, pp. 261–394, 1889).

14. J.A. Young, R.A. Evans, and P.T. Tueller, "Plant Communities—Pristine and Grazed," pp. 187–215, in *Holocene Environmental Change in the Great Basin*, Robert Elston, ed., (Reno, Nevada: Nevada Arch. Survey, Res. Paper No. 6, University of Nevada, 1976).

15. R.E. Eckert, Jr., *Improvement of Mountain Meadows in Nevada*, (Reno, Nevada: Research Report 4400, Bureau of Land Management, U. S. Dept. of Interior, 1975).

16. D.A. Klebenow, "The Habitat Requirements of Sagegrouse and the Role of Fire in Management," *Proc. Annual Tall Timbers Fire Ecology Conference*, (Tallahassee, Florida: Tall Timbers Res. Sta., 1972). Lieutenant H.L. Abbott, *Explorations for a Railroad Route from the Sacramento Valley to the Columbia River*, (Washington, D.C.: A.O.P. Nicholson Printers, Ex Doc. No. 91, 33rd Congress, 2nd Session, House of Representatives, 1855). D.E. Savage, "The Relationship of Sagegrouse to Upland Meadows in Nevada," M.S. thesis, Univ. of Nevada, Reno, Nevada, 1968).

17. C.D. Marsh, A.B. Clauson, and H. Marsh, *Larkspur or Poison Weed*, (Washington, D.C.: GPO, U. S. Dept. of Agric., Farmer Bull. 988, 1934).

18. Information on desert bighorns from E.R. Hall, *Mammals of Nevada*, (Berkeley, California: University of California Press, 1946). The last bighorn sighted in Elko County was in the Ruby Mountains in 1921.

19. E.R. Hall, "Mammals of Nevada." Quote is from J.H. Simpson, *Report of Exploration across the Great Basin of the Territory of Utah for a Direct Wagon-route from Camp Floyd to Genoa, in Carson Valley in 1859*, (Washington, D.C.: GPO, U. S. Engineer Dept., 1876).

20. Adelaide Hawes, *The Valley of Tall Grass*, (Bruneau, Idaho: Private Printing, 1950).

CHAPTER
4

TEXAS CATTLE AND CATTLEMEN

To many, "cowboy" equals Texas. Millions of words in story, song, and history have been written about this connection. Although this book deals with the Intermountain area, Texas cattle and cattlemen must be explored to trace the roots of and give perspective to cattle in the cold desert.

In the late seventeenth and early eighteenth centuries, there were abortive attempts by the French and later the Spanish to establish colonies in Texas. It was at that time that Spanish cattle were introduced to the territory that was to become Texas. Spain established a system of missions in south Texas, but met savage resistance from the Indians. After several false starts, the Spanish finally achieved a measure of control. But each time their missions were abandoned, the cattle were left to run wild in the brushlands. Finally, the Texas Territory passed to Mexico after the revolution of 1821.[1]

American colonists were moving into the Mexican territory of Texas throughout the early 1800s. Many came from the South, knew how to ride and shoot, but had no concept of the Spanish system of the open ranging of livestock. When Stephen F. Austin proposed a code of law to the Mexican government to govern Texas, the Mexicans added two articles they felt necessary. One article dealt with the registration of brands; the other pertained to the disposition of strayed stock.[2]

Contributions of the Spanish to western American livestock have been stressed repeatedly, especially in the writings of Walter Prescott Webb and J. Frank Dobie. Recently the writings of Terry G. Jordan have stressed the role of American colonists from the South in the roots of the Texas livestock industry.[3] Both schools of thought have merit and are contributing factors. The role of livestock in the early settlement of the South had long been overlooked. Jordan is to be commended for bringing recognition to the southern stock raisers. But in an eagerness to upset tradition, some supporters of Jordan's ideas substantially discount the influence of the Spanish on ranching's evolution.

Literature of the southern livestock industry carries continuing references to cowpens and herding. Hogs were as numerous as cattle in

71

many southern livestock operations. The southern environment permitted year-round grazing, and encouraged winter grazing. But, on the ranges of the West there would be no pens for working cattle and no herding in the northern European style. The holding of cattle by mounted men on an open piece of ground (rodeo ground) while other mounted men sorted the stock was completely foreign to southern livestock production.

After the Civil War there was a growing demand for Texas cattle. The northern states had prospered during the War and their market needs were building. The huge crews building the new transcontinental railroad system required food at many points. The vast area acquired by the United States through the Louisiana Purchase and Mexican War was developing. Much of this land was suitable for stock production, and there was immediate demand for red meat in the Far West mining boom towns.[4]

The Longhorns originating from Spanish cattle were brought to the New World from Spain. The Retinto is doubtless the major breed of Spanish cattle represented in the Longhorn. But in the remnants of relatively pure Spanish bloodlines found in isolated regions of the western hemisphere, one recognizes several regional breeds of Old Spain.[5]

A composite of Frank Dobie's descriptions provides an accurate portrait of the Longhorns. "Longhorns were tucked up in the flanks with high shoulders so thin they sometimes cut hailstones. Their ribs were flat and their length frequently was so extended that the back swayed. Viewed from the side, its frame would fool one into overestimating its weight. A rear view showed cat hams, narrow hips, and a ridgepole backbone. The tails of longhorns often dragged the ground despite their racehorse legs. The colors of longhorns illustrate what they were, a free breeding, outcrossing population of free roaming animals."[6]

Longhorns were slow to develop, not reaching maximum weight until eight to ten years old. Steers from four to eight years old averaged 800 pounds. Animals ten years old might reach 1,000 pounds. There did occur the rare animal that reached 1,600 pounds. Longhorns were slow converters of feed to beef and poor producers of the heavier cuts. Because they were long and lean and lacked a paunch, Longhorns dressed out surprisingly well. Their beef was inferior and they had a high percentage of bone compared to muscle. By an Indiana man's estimate, "one could salt in its horns all the roasting beef an average Texas steer could carry."[7]

Although the Longhorn was a poor-quality animal, spectators at any cattle market in the 1860s could see more bad than good specimens of American cattle.[8] Yearling Longhorns were lucky to reach 300 to 350 pounds. Three-year-old steers might reach 500 pounds, and five-year-old steers 750 to 800 pounds. This was an extremely slow growth rate compared to English breeds.[9]

Conditions under which the Texas longhorns were produced contributed to their quality. They received no supplemental feed, summer or winter. They roamed the grasslands and the woods for forage. The average winter death loss was 20 percent. Aside from quality of the beef produced, the inherent cruelty of the system was repugnant to many herdsmen. James MacDonald reported, "The prairies here and there are strewn with whitened skeletons; only an acclimatized Texan could contemplate with equanimity the fate of these unfortunate famished animals. At one side station more than fifty two-bushel bags full of bones were lying ready for transport."[10]

One thing could be said in behalf of the Longhorns—there were a lot of them. In the 1870s MacDonald calculated that America required eight brood cows per ten people to meet the annual consumption requirement for beef. In Texas there were over nine cows per person! The 1830 census estimated 100,000 head of cattle in Texas.[11] The 1850 census estimated 330,000, and that of 1860 estimated 3.5 million head.[12]

From the Texas revolution until the Civil War, cattle ran wild in Texas, multiplying rapidly. Sporadic attempts were made to market cattle to New Orleans, to California after the gold rush, and even to the North, but nothing was standardized. By 1837, cowboys began gathering herds of 300 to 1,000 wild unbranded cattle in the Nueces Valley, driving them to Texas cities to sell. In 1842, drives to New Orleans began. In 1842, there was a drive to Missouri. In 1846 Edward Piper drove cattle to Ohio for fattening. From 1846 to 1861 the drives increased. In 1850, drives went to California; in 1856, to Chicago. Until the Civil War there was continuing, though irregular, movement of cattle out of Texas.[13]

In 1865, cattle in Texas could be bought for three to four dollars per head, but found few buyers. The same cattle in the northern markets would have brought thirty to fifty dollars. Many Texans made vigorous efforts to connect the four-dollar cow to the forty-dollar market. In fifteen years, five million cows were delivered to the northern market, twelve to fifteen hundred miles away.[14]

Sedalia, Missouri, was the closest railroad shipping point, but southwestern Kansas, southern Missouri, and northern Arkansas were hostile to Texas cattle. There was fear of Texas fever being transmitted by the Longhorns, ill feeling from the Civil War, and only frontier law. On the whole, the season of 1866 was disastrous for Texans who tried to drive cattle to Missouri.[15]

First to see the desirability of a point of contact for eastern buyers and Texas drovers was J.C. McCoy. His plan was "to establish at some accessible point a depot or market to which a Texas drover could bring his stock unmolested, and there, failing to find a buyer, go upon the public highways to any market he wished." It was to establish such a market where the southern drover and the northern buyer could meet upon equal footing, undisturbed by mobs or thieves, that McCoy established the original "cow town" of the West—Abilene, Kansas.[16]

At times there was no market. Surplus cattle were held "on the prairie" or on ranches to be fattened. Thus, the cattle kingdom spread from Texas, utilizing the grasslands of the plains. Texas furnished the base stock, the supply, and the method of handling cattle on horseback. The plains offered free grass. According to Webb's classic assessment, "From these conditions and from these elements emerged the range and ranch cattle industry, perhaps the most unique and instinctive institution that America produced."[17]

Marketing of Texas cattle through westward-spreading Kansas towns continued through 1873. Until 1870 herds sent to Abilene and other railroads sold in a steady or rising market. Prices were particularly good in 1870, and movement from Texas in 1871 was the greatest in history—700,000 head went to Kansas alone. But late in 1871 market conditions changed, and drovers met a reversal. Half the cattle brought from Texas remained unsold and were wintered at a loss on the Kansas prairies. On September 18, 1873, the New York banking firm of Jay Cook and Company closed its doors, precipitating the first panic known to the range cattlemen. Thousands of Texas cattle were tanked for their horns and tallow with drovers receiving 1 to 1½ cents per pound.

J. Frank Dobie says, "After the Civil War all it took to become a cattleman in Texas was a rope, the nerve to use it, and a branding iron."[18] One of the more prominent stock drovers was Colonel D.H. Snyder, born in Mississippi, September 5, 1833. Moving to Texas in 1854, he settled in Williamson County. He bought Spanish horses in south Texas, drove them to Missouri where he traded them for draft horses. Building on his horse-trading experience, he delivered beef to the Confederate Army. His herds forded the largest rivers, including the Mississippi. He had three water-trained steers at the head of his herd. When these leaders took to the water without hesitation, other members of the herd followed safely.[19]

Colonel Snyder and his brother, J.W. Snyder, made one of their first early postwar cattle drives to the West, buying the cattle in Llano and Mason counties, Texas. They paid $1.50 per head for yearlings, $2.60 for two-year olds, $4.00 for cows and three-year olds, $7.00 for beef steers. They bought on credit, with notes payable in gold coins. They drove them across west Texas on the trail pioneered by John Chisum. Arriving at Fort Union, New Mexico, the brothers sold the cattle for $35 a head. They continued to Trinidad, Colorado, and sold the remainder to Charles Goodnight at $7.00 per head without requiring tally.[20]

In 1866 the Snyder brothers drove stock from Llano County, Texas, to Abilene, Kansas. After losing 140 head to Indians in the Indian Territory, they filed a claim against the federal government and were eventually paid for the loss. Colonel Snyder abolished the practice of slaughtering calves during long drives. On one of his initial northwest drives, after a day spent in bloody slaughter of calves of the herd, he issued orders that the slaughter be stopped. His herd arrived at its destination in good condition. A fair profit was realized from the calves, which weathered the waterless waste as well as the mature stock.

Colonel Snyder ran a taut ship, and possibly his greatest contribution to ranching was the training he gave young Texans in organization and leadership. The Snyder brothers had three rules of conduct for their cowboys: (a) you could not drink whiskey while in their employment; (b) you could not play cards and gamble while working for them; (c) you could not curse and swear in their camps or in their presence. One of their bright young cowboys, John Sparks, may have found some of those rules difficult to follow.[21]

The Snyder brothers changed tactics in 1870 by driving a large herd to Schuyler, Nebraska, seventy-six miles west of Omaha. This was the first herd to cross the Kansas-Pacific Railroad and continue to the Union Pacific. In 1871, they delivered four large herds to the Platte River Valley to establish their northwest headquarters at Cheyenne, Wyoming. The cattlemen of the plains were continually on the lookout for new range and ranching areas.

In 1872 the Snyder brothers delivered a large herd to a Texan in the far northwest corner of Utah on Goose Creek. The Texan was John Tinnin, an ex-confederate "Colonel" doing business in Utah as Ingram Company. Colonel Snyder selected John Sparks to lead this drive across the mountains to the Great Basin because Sparks had visited the area in 1868 while working for Colonel John J. Meyers. John Sparks was to spend most of a decade east of the Rockies before he again returned to the sagebrush/grasslands, but these early trips introduced him to the area. Meyers was termed the "Father of Israel" by drovers because he drove cattle to the wilderness. Records are not available, but most likely the Ingram Company was a livestock commission company and the cattle delivered to Goose Creek were for Jasper Harrell.

The Snyder brothers' biggest drive was in 1873, but upon reaching Wyoming they found that financial panic had destroyed the market. They borrowed money at 36 percent per annum to hold the herd through the winter. Part was then sold to a firm in Salt Lake City, so the brothers shod work oxen, bought fresh horses, and took the cattle across the mountains. Congress demonetized silver, the banks were closed, and their Utah buyers could not fulfill contract. Turning north to the Fort Hall Indian Reservation, Colonel Snyder entered bids to supply beef for the Indian reservation. He entered bids for each brother, half-brother, and brother-in-law. The army accepted one, putting the brothers in business on the sagebrush/grasslands of the Intermountain area. They delivered twenty-five beefs a week to the reservation, with the rest of their cattle ranging as far west as Rock Creek on the Snake River Plains.

The Snyder brothers contracted in 1877 to deliver cattle to J.W. Iliff, a cattle baron of Colorado; they then delivered 28,000 head of Texas stock. After Iliff's death, they were managers and administrators of his estate. They continued to supply Texas cattle to the Northwest until 1885, and were some of the first, last, and largest volume stock drovers. They were instrumental in supplying stocker cattle to establish the range livestock industry in northeastern Colorado, western Nebraska, Utah, and Wyoming. In 1885, Colonel Snyder was offered one million dollars for his Northwest interests. He refused, then lost most of his holdings in the hard winter of 1886 and the subsequent depressed markets. The Snyder brothers returned to Texas and became ranchers. The big days of the overland trail drover were past.

Other Texans did more than come and go; they transferred their cattle, their ranching, and their expertise to the sagebrush/grasslands. Perhaps the Texan who most influenced livestock production in the Intermountain area during the nineteenth century was John Sparks. What kind of man was he? John Clay, a Scottish-born livestock man who was widely respected in the American West, described Sparks, "He was tall, straight as a pine tree, with a clear flashing eye that seemed to look right through you. While somewhat deliberate in his movements, he was full of energy."[22] Clay concluded that Sparks had a wonderful knowledge of the cow business from his Texas background. Later in life, as a governor of Nevada, he was popularly known as "Honest John" Sparks. Sparks was a man who was noticed in a crowd. He carried what one could perceive as a John Wayne image, sporting English corduroy pants, western riding boots, a flask of whiskey, and a six-shot Colt revolver. In the winter he topped this with a coonskin coat.[23]

What kind of background did Sparks have that enabled him to develop a livestock empire? The roots of his family originated in England. The American portion of the family arrived in Maryland in the late 1600s, well before the Revolutionary War. John Sparks's grandfather, Millington Sparks, was a moderately successful planter in Maryland. John's father, Samuel Wyatt Sparks, apparently decided to move the family's endeavors to the promising New South.

John Sparks was born in Winston County, Mississippi, August 30, 1843, seventh of ten children.[24] His mother, Sarah, was a Deal from South Carolina. Later in life, John Sparks was described as having the manners of an aristocrat of the Old South. His was one of the early "new lands" families moving across the South with the forward edge of developing agriculture. The Sparkses were not plantation agriculturalists in the southern pattern of cotton and industrial crops. They developed new land, either brought it to production or "wore it out," sold their improvements, and moved on.[25] The family moved from Mississippi to Fountain Hill, Ashley County, Arkansas. Samuel was listed on the Arkansas 1850 census rolls as a farmer owning real estate valued at $1,200.[26]

The family continued its trek moving to Lampasas, Texas, settling about four miles east of town on Burleson Creek. This area of Texas borders on the blackland prairie soils which eventually were developed for cotton production. The census of 1860 shows Samuel Sparks owned real estate valued at $6,000 with improvements valued at $2,300. The family had done fairly well in the decade of the 1850s.[27]

When the Sparks family moved to Texas, John was fourteen years old. He later said he started in the cattle business at that point. He became proficient in riding, roping wild cattle in the brush, and following the Dobie axion, "a horse, saddle, rope, branding iron, and the guts to use them." He stayed with the family ranch until the outbreak of the Civil War when he was eighteen.

In view of the manpower needs of the Confederacy, compared to the more densely populated North, Sparks was destined to become a soldier. Texas was the only southern state that was also a frontier state. There were mixed emotions among Texans at the outbreak of the War, as the former republic strongly supported the concept of state's rights. Also, most settlers on the Texas frontier did not have slaves, nor did they necessarily want to fight for the institution of slavery. John Sparks may have fallen in this latter group. He found a compromise that allowed him to avoid combat against the North, but the option he selected was equally dangerous.

Ever since the Spanish first tried to establish missions in Texas, the Comanche Indians had offered stubborn resistance. Anglo settlers had pushed them across Texas in many bloody encounters. When federal troops withdrew with the start of the Civil War, the Indians were quick to take advantage. To cope, Governor Lubbock of Texas established the Frontier Regiments of Texas cavalry under the Texas Rangers, as an alternative to Confederate Army service. Many of the young Texans who later became leaders in the range livestock industry chose this form of military service. On January 29, 1862, just over a thousand Indian fighters were organized in nine companies of Texas Rangers under the command of Colonel James M. Norris. By spring they occupied eighteen outposts from Gainesville on the Red River to Fort Duncan on the Rio Grande.[28]

John Sparks enlisted as a private in Company 6, March 5, 1862, and was mustered into service at Keen's Ranch. Records indicate John Sparks received 40¢ per day for each of the three horses he furnished, and $5.47 for the arms, plus $12.00 per month salary. Each man carried two revolvers in his belt, two in saddle holsters, and a rifle in a scabbard. They trained by holding shooting matches and riding tournaments, with plenty of whiskey, women, and song. This was a rugged frontier action school for the young John Sparks.

In later years Sparks was often referred to as Captain John Sparks based on Civil War service. The limited records available indicate he went into the service as a private and was discharged as the same, serving from age nineteen to twenty-three. Considering the number of Texans who became "colonels" after the War, "captain" was a mild promotion.[29]

Sparks started back in the cattle business in Texas as soon as the Civil War ended. By 1868 he was driving and delivering Texas cattle to Wyoming. He got his first look at the Great Basin on a drive to northwest Utah while working for Colonel John Meyers. In those years, Sparks described himself as "employed by others and receiving a very small salary." He once delivered a large band of Longhorns from Texas to Virginia. He drove them first to Memphis, Tennessee, a trip of nearly 500 miles. They were loaded on cars of the Memphis and Charleston Railroad and shipped to Alexandria, Virginia. The steers were unloaded at Bristol, Tennessee, for forage and water. On the return trip, when Sparks stopped at Bristol, he was upset to find local cattle dying from Texas fever contracted from his stock's grazing area.[30]

John returned after his trip East, to Wyoming, and remained there two and a half years. In 1872 he returned to Texas and married Rachel Knight the daughter of Dr. D.J. Knight of Georgetown, Texas. During the same year, he and his brother Tom drove cattle to Wyoming and sold them at a good profit.[31]

Georgetown, Texas, was John Sparks's home and operating base for many years. This also was the Texas base for the Snyder brothers. In later years one of John Sparks's cowboys wrote an interesting letter to Dr. Fred W. Sparks, Professor of Mathematics, at Texas Tech. The cowboy and former wagon foreman, Tex Willis, wrote, "John Sparks did the same amount of work Charlie Goodnight did, except John made no noise. G.W. Littlefield, Ike Rogers, Ed Anderson, Ike T. Pryor all worked on the trail to Nevada with 'old John', as they called him and swore he was the greatest cowman they ever knew or worked for, but knew little about him. John was a top member of Texas Frontier Battalion in 1861, along with Charlie Goodnight. And in 1868, John Sparks got to California with his first 4,000 Texas cattle, and was paid off in an enormous amount of gold money for that day in time. Old John said simply, 'You have to trail cattle where people have gold money to pay for them.'"

In the early 1870s, John and his older brother Tom, "were engaged in the movement of four gigantic herds of cattle." Apparently all were profitable. Tom "was influenced by the rich Fort Hall bottomland to move to Idaho with his herd of 14,000." Tom was later wiped out in the big freeze of 1889–90.[32]

In 1873, John Sparks bought a large herd in Texas and drove them to Wyoming, establishing a ranch in the Chugwater River Valley. He established his wife and family in Cheyenne. The Chugwater River Valley in Wyoming is about seventy miles long and three miles wide. It covers an area of about 135,500 acres. The greater part of its length is bordered by high, rocky walls, and its fertile bottoms are quite low and level. The water supply is sufficient for irrigation of the natural meadows.[33] Hundreds of work cattle had been watered in the Chugwater River Valley as early as 1852 by Seth E. Ward, a trader.[34] Nelson Story trailed Texas cattle to Wyoming in 1866. The Union Pacific railroad reached Cheyenne in 1867; and, by fall of 1868, 300,000 Texas longhorns had reached Wyoming.[35]

Wyoming in the early 1870s was rapidly changing. The era of the American Indian and the bison was threatened by the introduction of settlers and cattle. The government seemed interested in rapid extinction of the bison to end the independent existence of the Plains Indians.[36]

Average annual precipitation at Cheyenne was roughly sixteen inches, compared to thirty inches in Williamson County, Texas. In both areas 70 percent of precipitation was received during the growing season.[37] Wyoming presented a more arid and much colder environment for extension of the range livestock industry from Texas. But aside from money to purchase cattle, investment in early Wyoming ranches was slight. A dugout cut into a hillside near a creek served as headquarters, with a similar dugout nearby for the horses.[38]

In 1873, the Chugwater River Valley was stocked with 4,100 head of cattle. Of them, 2,700 belonged to John Sparks. One might expect the young, recently married cowman to settle down and develop his ranch. But in 1874 Sparks sold his cattle and land claim to the Swan brothers. This was the first purchase by A.H. and Thomas Swan who were to assemble one of the largest land and cattle firms north of Texas. The purchase price was $35,000 and, in Sparks's words to H.H. Bancroft, "The largest sale ever made up to this time in the west." The Swan brothers paid $15,000 as a down payment. A chattel mortgage remained on the property until the balance was paid, an unusual arrangement at the time. How much land was included? Apparently no deeded acres were involved. Laramie County's General Index to Deeds and Mortgages 1867–1883 shows no record of the transaction. Federal Land Office Records in Cheyenne show no listing for John Sparks. Sparks may have established a preemption claim on the Chugwater, but he never filed the claim.[39]

Sparks joined the Wyoming Stock Grower's Association (WSGA) February 23, 1874, at its second meeting.[40] In a single decade this organization grew from ten members with 20,000 head of cattle valued at $350,000 to 435 members representing two million head of cattle worth $100 million dollars. The WSGA provided a means of organizing community roundups on the vast public lands; of protecting large stock owners from theft by registering brands and hiring brand inspectors; and, of developing organized political power in the territorial government.[41]

"In 1875 I purchased a herd in Colorado, drove them to Wyoming, and established a ranch on the North Platte River. This was the frontier ranch of that section of the country at the time. The Sioux Indians were so troublesome as to make it absolutely unsafe to go further into the interior."[42] Within a year, 1876, Sparks sold this property to Sturgis and Carre of Cheyenne. The agreement of sale in the Wyoming State Archives lists 3,000 head of cattle, plus ranch and horses. Then Sparks immediately purchased 5,000 head from the Swan brothers and located a new ranch on the North Platte River near Fort Fulterman. Sparks sold 2,700 head in 1874, and then bought 5,000 back in 1875 from the Swan brothers. He was pyramiding his capital by rapidly following the retreating Indian frontier, establishing ranches which he just as rapidly sold for profit. He continued this pattern in 1877, by selling his North Platte ranch to W.G. Irvine of Cheyenne.

Sparks made frequent trips from Texas to Wyoming to look after business interests, and combine his Texas and Wyoming activities. John and Bill Blocker delivered 3,000 head of cattle to a Sparks ranch west of Cheyenne in 1877. In 1878, Sparks purchased a ranch, range, and a large herd of cattle on Lodgepole Creek. This property was on the Wyoming and Nebraska territorial line. He sold it in the late 1870s to the Bay State Company. The Lodgepole Ranch became a very famous ranch during the cattle bonanza period of the 1880s.[43]

Sparks did not plow back all of his pyramiding capital into Wyoming ranches. Among other investments, he became a fifty percent partner with M.E. Steele in a bank in Georgetown, Texas. This bank was active in financing the developing cotton production industry in the Williamson County area. The railroad reached Georgetown in 1878, just as the area was coming to life after the long economic depression of Reconstruction. One of the major accounts of the Steele and Sparks Bank was D.H. and J.W. Snyder.[44]

Sparks also purchased 10,000 acres south of Taylor, Texas. There is no available record of the purchase price, but even at a minimum dollar per acre it was a substantial investment. He bought land from the Knight family and built a house on South Brushy Street. It was regarded as one of the finer homes in Georgetown. He continued to maintain this residence even after he was elected governor of Nevada, and often entertained the governor of Texas at the Georgetown home.

Personal tragedy struck Sparks in February of 1879 when his first wife Rachel Knight died at age twenty-six. They had two daughters; Maude born in 1874, and Rachel born in 1877, but who died at age three. On January 25, 1880, John Sparks married Rachel's half sister, Nancy Elnora Knight, with whom he had four sons, with three surviving infancy. Deal was born in 1880, Benton in 1882, Charles in 1885, and Leland in 1887.[45]

John Sparks was phenomenally successful in the cattle business throughout the 1870s, developing numerous Wyoming cattle ranches on the edge of the frontier. He was able to utilize his experience in a unique situation. It was a seller's market and John Sparks was selling his claims, range he did not own, primitive capital improvements, and low-cost cattle he had driven across the rugged trails and frontier miles from Texas. Eastern and foreign capital were begging for the opportunity to buy. Titled young men from Europe, and Ivy League graduates were rushing to Wyoming to invest in ranching, and for the chance to associate with those wild, young knights of the plains—the Texas cowboys.

Notes

1. C.W. Gordon, "Report on Cattle, Sheep and Swine," *Supplementary to Enumeration of Livestock on Farms in 1880*, 10th Census of the United States, Vol. III, (Washington, D.C.: GPO, 1880). For details of the Spanish influence, see also "The Spanish Approach to the Great Plains," in W.P. Webb, *The Great Plains*, (Boston, Massachusetts: Ginn & Co., 1931).

2. F.J. Dobie, *The Longhorns*, (Boston, Massachusetts: Little Brown & Co., 1941).

3. T.G. Jordan, *Trails to Texas*, (Lincoln, Nebraska: University of Nebraska Press, 1981).

4. J.G. McCoy, *Historic Sketches of the Cattle Trade of the West and Southwest*, Ralph Bicher, ed., (Glendale, California: The Arthur H. Clark Co., 1940).

5. J.E. Rouge, *The Cirello—Spanish Cattle in the Americas*, (Norman, Oklahoma: University of Oklahoma Press, 1977).

6. Dobie, "Longhorns."

7. Louis Pelzer, *The Cattleman's Frontier*, (Glendale, California: The Arthur H. Clark Co., 1936).

8. L.F. Allen, *American Cattle—Their Breeding and Management*, (New York: Orange Judd Co., 1890).

9. Gordon, "Report on Cattle, Sheep and Swine."

10. James MacDonald, *Food from the Far West*, (London: William P. Nimmo, 1978).

11. Ibid.

12. Webb, "Great Plains."

13. Gordon, "Report on Cattle, Sheep and Swine."

14. Webb, "Great Plains."

15. McCoy, "Historic Sketches of the Cattle Trade."

16. Webb, "Great Plains."

17. Ibid.

18. Dobie, "The Longhorns."

19. Snyder manuscripts on file in collection at Barker Historical Center, University of Texas Library, Austin, Texas. Unindexed material researched by Mrs. Joe B. Gordon.

20. Ibid.

21. Ibid. Colonel Dudley H. Snyder died on September 19, 1921, at the age of 88. He was a supporter of Southwestern University in Georgetown, Texas.

22. John Clay, My Life on the Range, (Chicago, Illinois: Privately Printed, 1923).

23. C.S. Scarbrough, Land of Good Water, Takachue Pouetsu, A Williamson County, Texas History, (Georgetown, Texas: Williamson County Son Publishers, 1974).

24. Obituary file for John Sparks, Nevada Historical Society, Reno, Nevada.

25. R.I. Fulton, "Camp Life on Great Cattle Ranges in Northern Nevada," Sunset, (1900) Vol. V, No. 3, pp. 111–118.

26. Seventh Census of the United States, Records for White Township, Ashley County, Arkansas, (Washington, D.C.: GPO, 1850). Sparks never mentioned the family move to Arkansas even though at least one of his sisters was born in White Township. Mrs. Joe B. Gordon traced the Sparks family from Mississippi to Arkansas.

27. Eighth Census of the United States, Records of Lampasas County, Texas, (Washington, D.C.: GPO, 1860).

28. J.B. Barry, The Days of Buck Barry in Texas, 1845–1906, J.K. Greer, ed., (Austin, Texas: Archives, University of Texas Library, 1936). Also see R.P. Felgar, "Texas in the War for Southern Independence, 1861–1865," Ph.D. diss., University of Texas, Austin, Texas, 1935.

29. Information on Sparks's enlistment from Civil War records, University of Texas, Austin. Research by Mrs. Joe B. Gordon. See also Prose and Poetry of the Livestock Industry of the United States; prepared by National Livestock Association, (Denver, Colorado: National Livestock Historical Association, 1904). The military service of John Sparks has been subject to a great deal of distortion. See D.H. Grover, Diamondfield Jack, (Reno, Nevada: University of Nevada Press, 1968). Grover even assigned Sparks to the Union Army.

30. John Sparks's personal history dictated to Hubert Bancroft, Bancroft Oral History Program, Bancroft Library, University of California, Berkeley, California.

31. Ibid.

32. J.H. Triggs, *History of Cheyenne and Northern Wyoming*, (Omaha, Nebraska: Herald Steam Book and Job Printing House, 1876).

33. A.W. Spring, *Seventy Years—A Panoramic History of the Wyoming Stock Grower's Association*, (Wyoming Stock Grower's Association, Laramie, Wyoming, 1942).

34. Maurice Frink, *Cow Country Cavalcade*, (Denver, Colorado: Old West Publishing Co., 1954).

35. E.S. Osgood, *The Day of the Cattleman*, (Chicago, Illinois: The University of Chicago Press, 1929).

36. Ibid. Quoting statements of Representative James A. Garfield in Congressional Record, 43rd Congress, Session I, pp. 2107–09.

37. *Climate and Man*, (Washington, D.C.: GPO, Yearbook of Agriculture, U. S. Dept. of Agric., 1941).

38. C.M. Owens, "Early Cattle Raising in Wyoming," M.A. thesis, Colorado State Teachers College, Greeley, Colorado, 1932.

39. Sparks's personal history dictated to Bancroft. See also Wilkerson manuscript of unpublished biographical material at Wyoming State Archives and Historical Department, Laramie, Wyoming.

40. Owens, "Early Cattle Raising in Wyoming."

41. Maurice Frink, W.T. Jackson, and A.W. Spring, *When Grass Was King*, (Boulder, Colorado: University of Colorado Press, 1956).

42. Sparks's personal history dictated to Bancroft.

43. M.S. Yost, *The Call of the Range—The Story of the Nebraska Stock Grower's Association*, (Denver, Colorado: Sage Books, 1966).

44. Scarbrough, "Land of Good Water." An account sheet showing the Snyder brothers' balance in the Steele and Sparks Bank is preserved in the Snyder Papers, Barker Historical Center, University of Texas Library, Austin, Texas.

45. Ibid.

II
THE
LAND
ACQUIRED

THE LAND ACQUIRED

Thermal winds swept up in the spreading alluvial fans rippling the endless strands of wheatgrass, Indian ricegrass, and needlegrass. Meandering stream channels supported sod-covered meadows in the canyon bottoms. The silver-gray ocean of sagebrush contained islands rich in potential for grazing animals. There were men of courage ready to risk fortunes in a grand experiment to determine if the cold deserts of the Great Basin could support a society based on extensive herds of cattle.

CHAPTER 5

BUY, BEG, BORROW, OR STEAL A RANCH

When John Sparks and John Tinnin bought an empire from Jasper Harrell, what did they actually own? Essentially they bought Harrell's claim to specific meadowlands along streams. The right to use the rangeland was unspecified by law. It was partially dependent on Sparks's and Tinnin's ability to defend its exclusive use.

A post-Civil War entrepreneur moving West to begin development of a stock ranch lacked sufficient capital to acquire land and livestock. The approach was to buy cattle cheap and use range free of charge. The vast bulk of the western states was owned by the federal government. Exceptions were the grants given railroads to facilitate construction, and grants to states when they entered the union. Grants were also made to public schools, universities, and other public works. Both railroads and states offered land received in grants for sale, under a variety of options.

The prospective rancher had ten options for acquiring public land. The options ranged from general practices to obscure processes (Table 1).[1] The first two options dealt with distribution of land through public auction. Following a proclamation by the President, or notice by the General Land Office, public lands were opened to auction. This was the usual procedure, as the frontier advanced to mid-America. Lands not sold remained "for sale" by the General Land Office.

A third method was the preemption claim, or "squatter's right." Settlers could improve 160 acres of unappropriated public lands, then later buy it at $1.25 per acre without competition. The preemption right was established by construction of a house and improvements. The settler had to file a declaration of intent to purchase within three months after settlement, or within three months of filing by the General Land Office on previously unsurveyed land. Preemption claims were often far in advance of land surveys. Due to delays in obtaining surveys, many preemption claims were held for years without final payment. Payment could be made up to eighteen months after filing declaration. Payment could be made in that rare commodity of the frontier, cash, or military-bounty land warrant, or agricultural college script.

Table 1. Methods of obtaining land from public domain during the late nineteenth century in the far western United States.

---

1. Public auction by General Land Office
2. Land offered for sale at auction, but unsold, remained available
3. Preemption claim
4. Homestead
5. Timber Culture Act
6. Timber and Stone Act
7. Desert Land Act
8. Mining Law
9. Coal Land Law
10. Military-bounty land warrant

A fourth method of obtaining public land was through homesteading. President Lincoln signed the Homestead Act on May 20, 1862, marking a reversal of earlier land policy. Before that date, the government held the Hamiltonian notion that it was the responsibility of the public domain to develop revenue and provide homes indirectly to settlers or pioneers. The Homestead Act marks adoption of the idea that the interest of government might be served best by providing land to settlers, later receiving compensation from increased national prosperity and from property values. This could then be used as the basis for public revenues.[2]

Provisions of the Homestead Act permitted settlers to acquire 160 acres free of charge, except for filing fees. The homesteader had to live on the claim for five years to receive title. The major differences between homesteading and preemption were the $1.25 per acre payment required under preemption, and five-year residency required on homesteads. An individual could combine the two methods to receive 320 acres of public land.

The Homestead Act reversed a longstanding policy of reducing the acreage of public land sales. Farmers were not interested in buying more land than a family could operate. The eastern United States had seen a steady decline in the size of auction blocks, from 640 acres down to 80. Prospective farmers had no objection to 160 free acres.[3] But to a congressman from the East, the grant of 160 acres of public land to a settler seemed a big giveaway.

Unfortunately, environmental conditions of the West required much more than 160 acres to support a livestock operation. For example, a hypothetical ranch providing 1,250 pounds per acre of annual herbage shows why 160 acres is inadequate. This herbage is produced from April through August with 80 percent usable by grazing animals. Each cow needs twelve acres per year (1,250 lb. herbage production x .8 forage utilization = 1 Animal Unit Month (AUM) per acre x 12 months = 12 acres). This gives a stocking capacity of thirteen cows per homestead, but a ranch could not run just thirteen cows. The herd must have a bull, and two replacement heifers; and, steers are not marketable until three years of age. With a 100 percent calf crop, the herd would consist of 4 cows (4 AUMs), 1 bull (1.5 AUMs), 2 replacement heifers (2 AUMs), 2 yearling steers (1.5 AUMs), 2 two-year-old steers (2 AUMs), and 2 three-year-old steers (2 AUMs), for a total of 13 animals. The two marketable steers have a value of $20.00 each after three years. So for the first three years of the five-year requirement, there would be a return of $13.33 per year. This hypothetical homestead collapses because the sagebrush/grasslands communities were not available for yearlong grazing. The 160-acre homestead was an economic and biological impossibility.

Homestead claims were used to acquire title to land that could be irrigated. If 100 of the 160 acres claimed were irrigable, the homesteader might produce eighty tons of hay, sufficient to winter eighty brood cows. But in their annual migration from winter ranges through the foothills to the summer ranges on the mountains, the homesteaders' eighty-cow herd required some 1,960 AUMs besides the hay produced by irrigating. Part had to be in the desert and part in the mountains to satisfy seasonal forage requirements. Both areas contained extensive areas of salt flats, rock outcroppings, and other nongrazable areas. In seasonal migration, a homesteader's eighty-cow herd might have to graze over 10,000 to 15,000 acres for sufficient forage.

A fifth method of obtaining land from the public domain, the Timber Culture Act, was possibly the most bizarre. Based on assumption that trees planted on the Great Plains would increase rainfall and make dryland farming practical, Congress passed the Timber Culture Law. This law granted 160 acres to the settler who could establish 675 trees on ten acres of the quarter section. The basic assumption was scientifically false and land entries under this law were often fraudulent.

A sixth method was under provisions of the Timber and Stone Act of June 8, 1878. One could obtain title to 160 acres of land unfit for cultivation, but valuable for the production of stone or timber, by paying $2.50 per acre. This was originally limited to California, Oregon, Nevada, Washington, and was little used in Nevada.

The seventh method of acquiring public land was the Desert Land Act, which permitted entry provided the land could be irrigated. The Act was never popular in Nevada. By the time it was passed, available irrigation waters were already appropriated.

An eighth method of obtaining land was through the mining laws. Secondary ownership after mining was finished could be of great strategic value if the land contained stock water. An individual could locate a lode claim 600 by 1,500 feet at $5.00 per acre, or a placer claim not exceeding 160 acres. These claims did not necessarily lead to a land patent. A patent could be applied for if the mineral claim was productive. The Elko newspaper often published notices in which John Sparks or Jasper Harrell were applying for patents on various mining claims.

The ninth method of obtaining public land was under the Coal Land Law, and related to the development of the western railroad network. The railroads were eager to develop local sources of energy in the West to avoid shipping costs. This method of land entry was important in Wyoming because of abundant coal deposits. Over 100,000 acres were patented in Wyoming under this act. This method of entry was not so important in the Intermountain area due to lack of coal deposits along the railroad, and only 1,600 acres of land were patented under this law in Nevada.

Land entry could also be obtained by using military-bounty land warrants. The warrants were awarded by Congress to soldiers for service in past wars. The warrants could be used to obtain land or sold to land speculators for cash.

So which method of land entry was most important in building the livestock ranches of the Intermountain area and specifically in the Great Basin? The surprising answer is none. The legal, primary methods of obtaining patents to public lands were not satisfactory for building ranches because of acreage restrictions. Every contemporary review of federal policies made from 1870 to 1900 confirmed this fact, but to no avail.[4]

Probably the most famous of the public land reviews during the late nineteenth century were made by Major James Wesley Powell. Among his recommendations, "The grasses of the pasturage lands are scant, and the lands are of value only in large quantities. The farm unit should not be less than 2,560 acres, the pasturage lands need small tracts of irrigable lands, hence the small streams of the general drainage system and the lone springs and streams should be preserved for such pasturage farms, the pasturage lands will not usually be fenced, and herds must roam in common. As pasturage lands should have waterfronts and irrigable tracts, and as residences should be grouped, as the lands cannot be economically fenced and must be kept in common, local communal regulations or cooperation is necessary."[5] Powell felt existing land laws were inadequate for settlement of irrigable lands.

Powell's report captured the inherent nature of the sagebrush/grasslands environment—irrigable lands and rangelands had to be tied together. Settlers needed small blocks of irrigated land to raise forage for wintering stock, plus extensive blocks of rangeland. The 2,560-acre blocks of range proposed by Powell were still much too small for the more arid areas, but the proposal was a step in the right direction. One of Powell's more radical proposals was abandonment of the rectangular system of land survey to allow land claims to fit soils and topography of specific situations. In irregular shapes, the eighty-acre irrigable tracts could fit available alluvial soils along streams.

Thousand Springs Valley, which constituted a major portion of the Sparks-Tinnin ranches, provides a classic example of the stringing of 40-, 80-, and 160-acre pieces of land astride alluvial and irrigable soils. Thousand Springs Creek starts and ends in the checkerboard of the Central Pacific Railroad grant. As the creek swings north in a great arc around Tony Mountain, it passes out of the checkerboard. In this area the private land occurs as a narrow band along the stream. The edges of the band are stairsteps caused by the joining of angular pieces of rectangularly surveyed land. Had Powell's suggestion been followed, a smooth band along streams could have been obtained.

Ranches were built not by direct entry on the public domain, but indirectly through the purchase of state lands. Over one-half of the deeded land in Nevada was obtained through purchase of state school lands.[6] When Nevada became a state in 1864, it received a number of land grants. These included 3.9 million acres for school support. After 1848, every state entering the Union received two sections (16 and 36) in each township for school support. An internal improvement grant of one-half million acres was the second biggest grant, followed by 90,000 acres for an agriculture college, 46,080 acres for public buildings, and 9,228 acres as an indemnity grant.[7]

Initial demands for state land were not met from the 700,000 unspecified acres ceded by the federal government in the statehood settlement. By 1871, timber, ranching, and farming withdrawals had depleted most of the unspecified acres, leaving the specific sections 16 and 36 of state school land grants in each township. In 1873, the state legislature of Nevada asked Congress to exchange this grant for one million unspecified acres, pointing out that sections 16 and 36 often occurred in the middle of barren playas or on the top of rugged mountains. In 1879 the Nevada legislature again approached Congress, asking for 1.5 million acres in exchange for the original sections 16 and 36 grant. This exchange was finally implemented on June 16, 1880, when a generous Congress authorized the transfer of 2 million acres of unspecified land to Nevada and accepted the unsold acres of the original public school grant in return. Of 3.9 million acres from sections 16 and 36, slightly over 63,000 acres had been sold.[8]

The two-million-acre grant, sold to applicants in maximum units of 640 acres, was depleted in less than twenty years. The bulk of these state school lands were selected in Elko, Humboldt, Lincoln, and Washoe counties, where there were large ranching companies. John Sparks became a major purchaser.

Between 1883 and 1893 state lands sold in Elko County totaled 367,926 acres. Between 1893 and 1903, 230,808 acres were sold, far above total acreage for any other county in Nevada. Most state school lands were sold for a down payment of twenty-five cents per acre, with the entrant either meeting the credit provisions of the purchase contract or forfeiting the acreage. Acreage restrictions, land prices, and interest rates were eased by successive legislatures to make the land acquisition process less cumbersome and expensive. Ranchers and speculators often held the land for decades without a single payment, while the state surveyor general and the legislature silently permitted the practice. By 1902, only defaulted lands, often overgrazed and stripped of usable timber, were available to prospective buyers of state school lands.

From congressional land grants in 1862 and 1864, the Central Pacific Railroad received five million acres of public lands in Nevada.

The railroad lands were checkerboarded on the right-of-way which paralleled the Truckee and Humboldt rivers over much of their length. These lands included a high proportion of the finest agricultural lands in the state.[9] The grant of public lands to the Central Pacific was more than double the state's grant for support of its schools. Little concern was expressed in Nevada newspapers or public documents over the magnitude of the railroad grants until the middle 1860s.[10] The movement of stockmen into Nevada accelerated after 1865, when California cattle buyers first contracted for Nevada beef.[11]

Settlers within Central Pacific's grant had the option of filing preemption or homestead claims under federal law, and/or filing for 320 acres under the State Land Act. The Registrar of the United States Land Office, Carson City, visited settlements along the Humboldt to accept entries locally.[12]

The Nevada State Board of Regents was desperately trying to support an embryonic school system during the 1870s. The only way to raise money was the sale of grant lands. The Board of Regents set a minimum figure of $1.25 per acre except for those lands in the railroad checkerboard which were priced at $2.50 per acre. The state attorney general found that land application could be accepted without competitive bidding, and opened 700,000 acres of unspecified state lands to entry without advertisement of competition. In only five years, all state lands, except the 16th and 36th section in each township, were sold. By 1874 most of the 700,000 plus acres of unspecified lands were sold.[13]

The 1880 congressional grant of two million unspecified acres enabled cattlemen to expand their land holdings at a time when both demand for beef and prices were rising.[14] This was significant for Nevada because mining output of the Comstock Lode declined after 1876, and impacted the state's economy. In 1881, the Nevada legislature authorized applications for 320 acres from the state grants of 1864, and up to 640 acres from the 1880 grants. The 1885 legislature rewrote the basic land statute, reducing the interest rates from 10 to 16 percent, and extending repayment from nine to twenty-five years. This was later extended to fifty years with 6 percent interest on the balance. Ranchers obtained use of the land for 6½¢ per acre per year![15]

According to historian John Townley, sales provisions enacted by the 1881 and 1885 legislatures were utilized to concentrate land among ranchers, with the State Land Office as an active participant. Assistance given individual applicants by land office staff violated both the letter and intent of the land statutes, but accurately reflected the attitude of the office during the twenty years required to disperse the two-million-acre school land grant. Applicants exceeding their legal limitation were openly advised to "have some member of your family who has not exhausted his or her right, or some person who will deed it to you, sign and return it to this office."[16]

Many of the largest land purchasers in Nevada, such as John Sparks, John G. Taylor, Miller and Lux, W.N.M. McGill, and the Dangbergs, deposited funds in Carson City to be drawn upon to meet the numerous annual credit payments for state school grant lands. These firms had dozens of contracts on various parcels. The State Land Office personnel kept track of the required payments for large ranchers in return for a private fee.[17]

The private lands that John Sparks and Jasper Harrell acquired generally were restricted to areas that could be irrigated or to strategic springs. These areas were often rather narrow, as the land along the middle portion of Thousand Springs Valley. Near San Jacinto where Trout and Shoshone creeks joined Salmon Falls to form a broad valley, Sparks and Harrell acquired all the available bottomland that could be irrigated.

The major exception to purchasing only irrigable land was John Sparks's acquisition of Gollaher Mountain. Located east of San Jacinto, Gollaher Mountain's 8,000-foot highland forms the divide between streams draining to Salmon Falls Creek and Rock Creek to the south and Goose Creek to the east. In a series of acquisitions from 1890 to 1896, which would provide a study of nineteenth century land policy, John Sparks acquired title to nearly a full township of high-elevation rangeland on Gollaher Mountain. This is one of the rare examples of a nineteenth-century rancher obtaining title to his summer rangelands. It is also surprising that he initiated these extensive purchases in 1890, a year when Nevada cattlemen were extremely short of cash.

After 1880 Central Pacific Railroad offered land at 20 percent down and 7 percent interest over five years. Prices ranged from two to ten dollars per acre. Central Pacific held the land without patent from the federal government and untaxed by local governments. In 1885, Central Pacific's holdings were divided into natural ranges based on topography, with each range offered for lease as a whole. Some were as small as

10,000 acres and some as large as 250,000 acres. The largest of all was the 500,000 acre lease on Thousand Springs Creek which went to Sparks and Tinnin. The railroad leases were usually made at 2½ cents per acre.[18]

The Nevada legislature on March 11, 1889, required all preemptive rights to be recorded by December 12, 1889. A second statute made owners of livestock trespassing on private property liable to double damages.[19] The first bill affected small landowners who had squatted on small water sources for grazing since the territorial period. Most had protected their improvements by preemptive claim, but had never applied to purchase the land from either the state or the federal government. Now they were forced to make entry or pay for the land in full. Many small ranchers simply did not have the capital necessary to purchase the land on which they were squatting. Their only recourse was to sell to the larger ranchers. The second law allowed ranchers who owned watering points to refuse to allow livestock other than their own to use the water.[20] The trespass law strengthened the control of the ranchers with deeded land over migratory sheep, but the fact that sheep could use snow for water on winter range negated the law.

From 1887 to 1899, land statutes remained substantially intact. Terms were liberal, repayment provisions rarely enforced. Annual land entries declined from a peak of 400,000 acres to an average of 60,000 acres.[21] This period was marked by both a national depression in the early 1890s plus prolonged depression in Nevada.

In 1899, the legislature finally asserted itself by requiring prompt payment on contracts. The State Land Office could receive overdue payments one year from the due date without the entrant's losing the land. But after that period, contracts in arrears were made null and void. This resulted in voluntary forfeiture of hundreds of thousands of acres. By 1903, nearly a million acres had reverted.[22] In March, 1900, the State Land Office advised each county that the school land grant was closed. The preceding twenty years had seen a period of economic depression for Nevada, but the two million acres of state school grant land had passed into private hands to found a vigorous cattle industry.

Ranchers often acquired land by using an intermediary for the actual entry. Cowboys were hired to enter on land, and for small sums, turn it over to the employer. Or, the cowboy homesteaded the public land, then applied to have the homestead converted to a preemption

97

claim. After conversion the rancher paid the $1.25 cost of the land, and a fee to the cowboy who transferred title to the rancher.[23]

The Homestead Law made no provision for the government's taking back land on which the settler had failed or become disillusioned after receiving patent. Ranchers often obtained land from banks and stores that had lent money to homesteaders, then received the homestead when the settler decided or was forced to leave.[24]

Nevada represents the ultimate in federal ownership of lands among the adjacent forty-eight states. From the original cessation until 1934 when President Roosevelt closed remaining vacant lands to entry, only 6 percent of available public domain, excluding railroad grants, passed into private ownership. There was often no legal way to obtain title to the acres of rangeland necessary to sustain livestock. Even if there had been, the ranch operations probably could not have survived the tax burden that ownership of such lands would have imposed.[25] The only option was to use the public lands, and try to protect possessory right to grazing those lands. John Sparks used the years from 1880 to 1900 to give his empire legal substance in terms of land ownership. Despite the fact that he owned only a fraction of the total rangeland he used, he gained ownership of irrigable lands where hay could be produced, and parlayed the package into a western empire.

Notes

1. B.H. Hibbard, *A History of Public Land Policies*, (New York: The MacMillan Co., 1924).

2. *United States Land Office Report, 1875–1876*, (Washington, D.C.: GPO, No. 1680, 1876).

3. W.P. Webb, *The Great Plains*, (Boston, Massachusetts: Ginn & Co., 1931).

4. Gordon, "Report on Cattle, Sheep and Swine," *Supplementary to Enumeration of Livestock on Farms in 1880*, 10th Census of the United States, (Washington, D.C.: GPO, 1880). See also Thomas Donaldson, *The Public Domain, Its History with Statistics*, (1970 ed., Johnson Reprint Corp.; original, Washington, D.C.: GPO, 1889). Also see the report from the Chief of the Bureau of Statistics (Nimmo Report) in response to a resolution of the House calling for information in regard to the range and ranch cattle industry in the

western United States and territories, (Washington, D.C.: GPO, Ex. Doc. No. 267, 48th Congress, 2nd Session, House of Representatives, 1885).

5. Major J.W. Powell, *Report on the Lands of the Arid Regions of the United States*, (Washington, D.C.: GPO, 1878).

6. E.O. Wooton, *The Public Domain of Nevada and Factors Affecting its Use*, (Washington, D.C.: GPO, Tech. Bull. No. 301, U. S. Dept. of Agric., 1932).

7. Based on data prepared by Hugh A. Shamberger, former Nevada state engineer and held in Shamberger Collection, Nevada State Historical Society. Summary of data presented by J.M. Townley, "Reclamation in Nevada, 1850–1904," Ph.D. diss., University of Nevada, Reno, Nevada, 1976.

8. Townley, "Reclamation in Nevada."

9. George Kraus, *High Road to Promontory*, (Palo Alto, California: American West Publ. Co., 1969).

10. Townley, "Reclamation in Nevada."

11. John Townley cites the *Sacramento Daily Union*, 28 January 1865, p. 3, as one of the first references in California to Nevada being a source of beef.

12. U. S. Bureau of Land Management, Records Group 49, Office of the U. S. Surveyor-General for Nevada, Federal Record Center, San Bruno, California.

13. "Biennial Report of the Surveyor-General and State Land Register of the State of Nevada for 1867 and 1868," in the *Journal of the Senate*, 4th session of the legislature of the State of Nevada, 1869, begun on Monday, 4th of January and ended on Thursday, 4th of March, (Carson City, Nevada: State Printer, 1869).

14. Townley, "Reclamation in Nevada." The fixation with artesian water was not limited to Nevada. The U. S. Geological Survey conducted ground-water surveys in many western states and territories during the late nineteenth century to determine the potential for artesian wells.

15. Townley, "Reclamation in Nevada."

16. Ibid. Townley based most of this discussion on letters found in the U. S. Bureau of Land Management, Record Group 49, Office of the U. S. Surveyor-General for Nevada, Federal Record Center, San Bruno, California. Specifically, Townley quotes a letter from E.D. Kelley to Thomas Short, 21 March 1900; and a letter from E.D. Kelley to H.H. Porch, 10 March 1900.

17. Ibid. Townley based these charges on letters found in the U. S. Bureau of Land Management, Record Group 49, Office of the U. S. Surveyor-General for Nevada, Federal Record Center, San Bruno, California. Specifically, a letter from J.E. Jones to John Sparks, 14 January 1890; a letter from H.D. Notware to

Miller and Loy, 14 September 1891; a letter from A.C. Pratt to Chris Dansberg, 19 June 1896; and, a letter from A.C. Pratt to John G. Taylor, 25 June 1896.

18. L.W. Mills, *A Sagebrush Saga*, (Springville, Utah: Art City Publishing Co., 1954). Also see, Townley, "Reclamation in Nevada."

19. Statutes of the State of Nevada passed at the 14th session of the Legislature, began on Monday, 7th of January, and ended on Thursday, 7th of March, (Carson City, Nevada: State Printer, 1889).

20. Townley in "Reclamation in Nevada" cites the *Belmont Courier*, 10 August 1889, p. 3, for an article on the necessity for smaller ranchers to sell to larger ranchers.

21. Townley, "Reclamation in Nevada."

22. "Biennial Report of the Surveyor-General and State Land Register, 1901–1902," *in Appendix to Journals of Senate and Assembly*, 21st session of the legislature of the State of Nevada, (Carson City, Nevada: State Printer, 1903). See also, Statutes of the State of Nevada passed at the 19th session of the legislature, commenced on Monday, 16th of January and ended on Friday, 10th of March, (Carson City, Nevada, State Printer, 1899).

23. George Stewart, "The History of Range Use," pp. 119–33 *in The Western Range*, Senate Doc. No. 199, 74th Congress, 2nd session, (Washington, D.C.: GPO, 1936). See also Hibbard, "History of Public Lands," and E.S. Osgood, *The Day of the Cattleman*, (Chicago, Illinois: University of Chicago Press, 1929).

24. Maurice Frink, W.T. Jackson, and A.W. Spring, *When Grass was King*, (Boulder, Colorado: Univ. of Colo. Press, 1956), pp. 1–24.

25. Wooton, "Public Domain of Nevada."

CHAPTER
6

JOHN SPARKS: CAPITAL, CREDIT, AND COURAGE

John Sparks had a good thing going in Wyoming. Unfortunately there was a limit to the rangelands available east of the Rocky Mountains, and by the end of the 1870s, ranges were nearly fully stocked. Sparks did not come to the Intermountain area to exploit the virgin range. He came with ready capital, credit, and expertise. He brought appearance, poise, the ego of a cattle king, and an abundance of nerve.

As he later said to H.H. Bancroft, "In 1881 I went to Nevada and formed a partnership with the John Tinnins of Elko County. The range is known as the 1,000 Springs Valley Ranch and Range. In 1883 our firm purchased the Barley Harrell property, on the Salmon and Snake rivers in Nevada, and the territory of Idaho. The entire property now owned by this firm is known as the Rancho Grande. The great bulk of the stock is blood stock, being an admixture of Hereford and Shorthorn. The firm is a member of the National Stockgrower's Association. We carry on our business without a ledger of any kind; don't even have a bookkeeper. In fact, I keep my accounts in my head. We ought to and probably soon will have a bookkeeper. We now have between 80,000 and 90,000 head of cattle and are almost land poor, as the expression goes."[1]

In his new ventures of the 1880s, Sparks joined with John Tinnin. Colonel John Tinnin's name is frequently mentioned in connection with significant events, but concrete facts about his background are difficult to find. One source indicated he, too, was the son of a Mississippi planter, and that he fought Indians as a Texas Ranger during the Civil War.[2] In the late 1860s, Tinnin was employed as a livestock commission agent handling the sale and delivery of Texas cattle to the Intermountain area for the Ingram Company of Salt Lake City, Utah. Tinnin also owned a classic steamboat Gothic house at 1220 Austin Avenue in Georgetown, Texas. He bought the house about the same time John Sparks became well-established in that growing town. Tinnin's home was noted for its beautiful furniture and its parrot. The vocabulary of the parrot frequently shocked visitors.[3] Mrs. Harold G. Scoggins of Georgetown remembers, "It was an interesting old home.

When I was growing up here, it was furnished in rosewood furniture, even the piano. It had a circular stairway from the kitchen to the master bedroom. There was a square tower at the front—over the entrance. Next to my dad I loved Mr. Tinnin, in spite of the fact I was punished for using his 'cuss' words."[4]

Sparks and Tinnin purchased their first ranch in 1881; this was "Old Bill" Downing's H-D Ranch on Thousand Springs Creek. "Old Bill" had departed an immigrant train to found the H-D. Though not a large ranch, this was a strategic location for control of the upper reaches of Thousand Springs Valley. On November 6, 1881, the Elko Independent announced Sparks-Tinnin had purchased the Jasper Harrell and Armstrong property at Tecoma where Thousand Springs Creek flows north of Pilot Range onto the Bonneville Salt Flats. The property was valued at $150,000, but the purchase price was not announced. The sale did not affect Harrell's holdings on Salmon Falls and Goose Creek or his extensive holdings in Idaho.[5]

On June 15, 1883, the Elko Independent reported the sale of the remaining Jasper Harrell properties to Sparks-Tinnin. The sale price was reported as $900,000, with $100,000 down and the balance due in eight yearly installments of $100,000 each, with a 4 percent interest on the balance. The sale included 30,000 head of cattle, a large number of horses, and extensive rangeland said to be one hundred miles square. Harrell customarily branded 10,000 calves on this range and the previous year had shipped $120,000 worth of beef.[6]

Why did Jasper Harrell decide to sell holdings, and on such favorable terms? Was he tired of overseeing the widespread operation? He had accumulated enough wealth to last his lifetime. The wily old 49er may well have realized the ranges were overstocked and the beef bonanza was vulnerable. He sold out for a low down payment, with low interest on the balance. He had sent his son, Andrew Jasper (A.J.) Harrell, to Nevada in late 1870s to learn the cattle business. A.J. had graduated from Heald Business College in San Francisco. Did Jasper sell so he and his son could concentrate on banking and real estate interests in California?[7] The answers were probably many.

In November 1883 the Elko Independent printed an article with headlines calling Sparks-Tinnin "the cattle kings of the west." The 70,000 head of cattle they owned and 17,000 calves they branded annually made them the "largest ranchers in the west." The Swan brothers of

Wyoming ran more cattle, but it was a stock company rather than individual ownership. Judge Carey of Cheyenne, who ran 30,000 head under his individual ownership, was reportedly the "second largest operation." This article was reprinted by the *Weekly Drover's Journal*, Chicago, giving Sparks-Tinnin national recognition.[8]

In the 1880s money for ranch operations was hard to obtain and expensive to borrow. John Clay once had trouble getting a $5,000 operating loan when the ranch was valued at $500,000, debt-free, with 5,000 steers to market that fall.[9] Interest rates were quoted at 1 percent per month, and 15 percent per year. An eastern banker loaned money on Nevada cattle and came west to verify the collateral. He rode the range for days, covering several hundred miles in a buggy. He saw only a handful of cows representing collateral, and commented that he "might as well loan money on a school of fish in the Pacific Ocean."[10] During good years most cattlemen liked to count their money in cattle. When dry years and low prices occurred, they found it difficult to convert cattle into cash sufficient to cover expenses without impairing operations.[11]

Sparks-Tinnin had $100,000 plus interest to pay annually on a ranch operation that had grossed $120,000 in 1882, the highest cattle price year of the decade.[12] It is possible that they had inside knowledge that Jasper Harrell actually had more cattle than he realized. C.W. Hodgson considered this a classic case of the so-called "book sale."[13] Ranches were sold based on the number of head carried in their books. These numbers were often greatly inflated. This lead to bizarre incidents of Scottish accountants who painted identifying marks on cows in an attempt to balance the cows on the range with the number of cows in the books. In the case of Sparks-Tinnin it could have been a reverse "book sale," where the number was lower than the actual head. John Sparks was once asked how many cattle he owned. "We leave those matters to the assessor and he comes around once a year. It is an unwritten law that a cattleman never talks of the size of his herd."[14] During the course of his operations in Nevada, a variety of cattle numbers were attributed to his ownership. John may have adjusted the number to suit the occasion as most ranchers did, but in truth, probably did not know the actual number.

A major factor that aided the growth of ranching in Nevada during the nineteenth century was a favorable political climate. One of the

better-known cowboy governors was Lewis Rice Bradley, the second
governor of Nevada, who served from 1871 to 1878.[15] Bradley drove
cattle from Missouri to Stockton, California, in 1852.[16] He led what was
known as the "Bull Block" in the state senate, because the group stood
for the interest of ranchers when Nevada politics were more concerned
with affairs of the Comstock miners. Bradley had ranched in California
until 1862 when an exceptionally dry winter caused him to turn to
Nevada sagebrush range. He established himself first in central Nevada
to furnish beef to the Austin mining district. He later moved from
central Nevada to Pine Valley, and the South Fork of the Humboldt
River in Elko County. The governor's son, J.R. Bradley, in partnership
with the Russell family, developed a huge ranching operation extending
into the Snake River Valley and lying just west of Sparks-Tinnin
holdings. They had common fall roundups or rodeos where the present-
day Twin Falls, Idaho, is located.[17] Nevada cowboy governors after
Bradley included Jewett W. Adams, 1881-1886, and Reinhold Sadler,
1896–1902, who also had large ranching interests.

The primary market for Nevada cattle during the 1880s was the
growing population of California. Relatively few packers controlled
much of the market. In the San Francisco area, Miller and Lux began to
dominate markets and have a major voice in establishing live-beef
prices. Henry Miller made application to the California State Legislature for a butcher's reservation in south San Francisco, and a guarantee
that butchering might be carried on for ninety-nine years. Miller
contracted for erection of a slaughterhouse wharf big enough to
accommodate every butcher in San Francisco. Eventually destroyed in
the 1906 earthquake, the supporting piles remained for years because of
the thousands of tons of offal dropped under the wharf.[18]

Self-sufficient in beef production by 1870, Nevada supplied an
estimated 30,000 animals annually to the California wholesale
butchers.[19] Later in the 1870s, a group of Reno and Winnemucca busi-
nessmen decided to butcher and ship their dressed beef to California.
Thirty carcasses, rather than eighteen to twenty live animals, could be
shipped per railroad car. California Wholesale Butchers' Association
broke the new Nevada company by refusing to sell to any retailer in
California who bought the Nevada-dressed beef.[20] By 1880 Nevada was
supplying one-half of San Francisco's beef supply. In 1884 it was

estimated that the San Francisco market required 250 head per day. Long trains of cattle cars filed through Reno loaded with Elko County cattle on the way to the Bay Area market.[21]

Freight rates from Halleck, near Elko, to Chicago were $260 per car, while the rate from Halleck to San Francisco was $120. The trip east was at least eleven days with feed and water necessary along the way.[22] With this differential in mind, the California Butchers' Association quoted lower prices than eastern markets. With a price drop in live cattle after 1885, the San Francisco market was glutted.

Nevada ranchers turned to eastern markets to find an outlet after 1885. John Clay became acquainted with Sparks-Tinnin in 1885, when he bought 3,000 steers from them delivered at Rawlins and Rock Creek, Wyoming. The 2-year-old steers cost $27; the 3-year olds $35.[23] The same year Sparks-Tinnin shipped 1,900 head of young mixed stock to J.G. Pratt, land agent for the Union Pacific Railroad, delivered at Antelope on the Cheyenne River.[24] In October 1887, the Elko paper reported Sparks-Tinnin had just returned from marketing cattle in Omaha.[25]

Shipping by rail cars was an innovation that made the beef bonanza possible. This was especially important in the Intermountain area. The region was rimmed with mountains, forests, and deserts which made it virtually impossible to walk fat animals to market. Despite their value, railroad stock cars were hard on cattle. Sudden starts and stops caused injury. If cattle went down, they were usually trampled by the other stock. Losses from Colorado or Wyoming points to Chicago were estimated at 1.5 per 1,000 shipped.[26] In 1872, the American Humane Education Society offered a $100 reward for the best essay on the transportation of animals. George T. Angell, writing under the pen name of Litera, won the prize and 20,000 copies were printed and distributed. In 1873 the twenty-eight-hour law was passed by Congress after prodding by the American Humane Education Society. This law required all stock to be removed from trains at least every twenty-eight hours for a period of five consecutive hours for feed, rest, and water. Stockmen, commissions, and meatpackers bitterly fought this law.[27]

During the 1860s it was customary to crowd thirty steers into a rail car. Federal regulations were passed which required one-third of the animals must have room to lie down. Acceptable carloads were based on live weight of the animals being shipped: 700-pound animals were twenty-three per car; 1000-pound animals were twenty per car; and 1400-pound beefs were sixteen per car. Ranchers bitterly protested this law, claiming it provided room for cattle to go down and be trampled, while the densely packed cars actually protected the animals.

The shipment of cattle was a special operation that stirred strong emotions in ranchers and cowboys. It was hard work done under supercharged conditions against specific deadlines. The railroad would only hold the cars on a siding for a short time without an additional surcharge. The rancher had to request cars and plead with the agent to have them spotted on the siding at the correct time. The steers had to arrive at the shipping corral in sequence with the cars. Wild range steers, half-broken saddle horses, locomotives belching clouds of steam, and shrill whistles were volatile components that did not mix well. A toot on the locomotive whistle at the appropriate time by a sly engineer would set off a rodeo that might end with several cowboys on the ground.

Shipping corrals were located in Elko, Carlin, Tecoma, Wells, Deeth, Halleck, Palisade, Red Rock, and Iron Point along the Central Pacific route through Elko County. The steers were held in bunches on open ground. Then, sufficient animals to fill the corrals were cut out and forced through the gate. Inside, a crush corral funneled steers into the loading chute. Corrals were either choking with dust perfumed with cow droppings or ankle-deep in sucking mud. The steers had to be forced up the chute with shouts, curses, or horses and cowboys poking poles through cracks in the chute to jab animals in the ribs.

The classic "wide horns and narrow chute" cartoon shows cowboys trying to force a steer with eight-foot horns through a four-foot opening rail car door.[28] If a dog barked at the wrong time, the steers would balk or try to turn in the chute. The range-raised steers would turn and fight if provoked. Men on foot in the corrals were in constant danger.

After the last car was loaded and rolling down the tracks, the cowboys were in town with the opportunity for a spree. In Elko, they headed for Dobe Row, a block and a half of adobe houses which formed a noted red-light and gambling district. The rancher could now relax with the thought that cash money was on the way from the sale of the shipped animals.

To determine the composition of range cattle herds, the Tenth

United States Census interviewed nineteen Nevada ranchers in 1880. The census occurred one year before Sparks-Tinnin came to Nevada, but it provides insight into range operations of that decade. The nineteen herds contained 94,786 head, broken down as follows:[29]

	Number	Percent of Herd
Bulls	1,422	1.5
Cows	28,439	30.0
Three-year-old steers and beefs	11,861	12.5
Two-year-old steers	14,219	15.0
Yearlings	19,907	21.0
Calves	18,948	20.0

It was custom during the 1880s to allow bulls to range throughout the year with other cattle. This led to the untimely dropping of many calves when severe storms often proved fatal to both cow and offspring. The percent that survived to yearlings was estimated at 66 to 80 percent in 1880 for Nevada. Estimated average annual loss among cattle over two months old was 6 percent arising from disease, winter and spring storms, snakebites, wild animals, theft, and poisonous weeds.[30] It is interesting that the ranchers who responded to the 1880 census failed to list starvation as a cause of death.

Theft was a problem. Colonel E.P. Hardesty, Elko County, gave instruction to his cowboys regarding anyone they caught killing his cattle, "If he stole it to eat, tell him to enjoy it and bring me the hide. If he stole it to sell, bring me his hide."[31] Sam McIntyre, who ranched off the North Fork of the Humboldt, ran Galloways, the Scottish Highland cattle. Because of the uniqueness of the breed, he did not bother to brand his stock, until theft became a serious problem. A reporter once asked Sam who else in Elko County was running Galloways and Sam replied, "everyone who has a horse."[32]

One Elko County ranch foreman was much less charitable. He caught a homesteader skinning a company cow. He forced the home-steader to nail the hide to his shack with the brand showing so that everyone who passed would know he was a thief.[33]

The 1880 census estimated "Cost of going into the ranching business on the Humboldt River in Nevada:"[34]

2000 three-year-old cows	@ $12.50	$25,000
100 two-year-old bulls	@ 50.00	5,000
25 saddle horses	@ 50.00	2,250
2 work horses	@ 100.00	200
1 wagon and harness		120
Ranch building, saddles, etc.		2,000
		$34,570

Annual expenses of a Humboldt River Ranch were listed as:

Five cowboys	$40/month for 8 months	$ 1,600
One cook	$30/month for 12 months	360
Two cowboys	$40/month for 2 months	320
Provisions for men	$12/month	720
Taxes on cattle		450
Taxes on horses		30
Taxes on ranch improvements		27
		$ 3,507

Experienced herdsmen estimated their annual profit on capital invested at 20 to 30 percent if death losses were not greater than 5 percent.[35]

In 1885 the average cost of raising a steer, including interest on capital invested, was estimated by the larger stockowners at 75¢ to $1.25 per year. Thus, a large four-year-old steer ready for market cost $4 to $5 to raise.[36] During the decade of the 1880s, prices stayed in the $15 to $20 range, leaving a considerable margin. When 1884 brought the second depression since the Civil War, the New York firm of Grant and Ward failed. And later, in 1866 and 1867, severe winters in Wyoming created panic in the livestock market.[37]

Analysis indicates that only one-third of the cows produced a calf each year. Often calves were left on the cows until they weaned themselves. The big sucker calves prevented the cows from coming into heat or conceiving. Some ranchers tried to improve the quality of their stock during the 1880s, but this was made difficult by the ranging of several operators on common range. The stockman who turned out superior bulls on the common ranges divided the benefits with all the brands.[38]

There are several references to the Sparks-Tinnin attempts to improve quality during the 1880s by castrating bull calves, introducing improved bulls, and reducing the numbers of stock on the range.[39] The reduction in stock number may well have been aimed toward making their mortgage payment rather than motivation to improve the range.

In view of many observers the ranges were being overgrazed. The editor of the Carson City, Nevada, *Morning Appeal* on December 4, 1886, called for "appropriation of state funds for research to find ways to seed and restore the range."[40] But it was half a century before large-scale restoration techniques were put into practice. In 1885, a special agent of the Bureau of Animal Husbandry reported, "Cattlemen are warning that the western ranges are overstocked and petitioning Congress to lease the then public domain at 1¢ per acre."[41]

The vocabulary of ranching in the western United States is largely borrowed from Spanish. But the most important things the Spanish contributed to range livestock operations were the techniques necessary to work cattle in the open without fences and corrals. The lariat, a type of saddle, chaps, and the sombrero came along with those techniques. The very newness as well as the immensity of the outfits left the Americans without standards by which to gauge either the security of the roaming herds, or the capacity of the forage to hold up under continued intense utilization.[42]

In 1880 the foremen of many of the large ranches in Nevada, eastern Oregon, and southern Idaho were called "major domo" rather than foreman. The cowboys were "vaqueros," about three for every thousand cows. The agrarian culture in California before the Mexican War was akin to the fifteenth-century Castile or Extremadura. The invasion of California by Anglo settlers, the breaking up of huge ranches, and the rise of dryland cereal production spelled the end of their way of life in their homeland. They came to the sagebrush/grasslands with Spanish cattle.

The use of rawhide for cowboys' tools was an important part of the Spanish heritage. The term "rawhide lariat" derives from the Spanish "la reata," and became a symbol of this craft technology.[43] To make a lariat, the pliable hide of a relatively young animal was preferred. The lariat was braided from four, or occasionally, eight strands. Maximum diameter was usually three-eighths of an inch. The fresh hide was allowed to dry for four or five days. If a dried hide was used, it was soaked in water, with wood ash added to loosen the hair and soften the hide. The cutting of the strands began in the center of the hide and worked outward. A sharp, thin-bladed knife was used, plus a wooden gauge for control of thickness. Strands for a seventy-foot rope had to be ninety feet long to allow for plaiting. Hair was stripped from the hide with a bit of broken glass. The braiding took many hours. The four strands were tied together, and the weaving began over one, under one, right and left, drawn tight with no slack. At the end the strands were worked back to form a loop or hondo.

After braiding, the lariat was stretched and rubbed with tallow. A heavy weight was often tied on and it was twisted and stretched until perfectly round. The lariat had to be kept dry, and it was not as strong as manila grass or sisal rope. The rawhide rope was not strong enough to stand hard and fast tying, so a roper could not tie the rope to the horn before roping an animal. Ropers had to take a dally, or twisting of the rope around the horn, and let out the slack. Because rawhide ropes could not be tied hard and fast, and dallies had to be taken, ropers were sometimes identified by their lack of thumbs. If a thumb got stuck in the dally when the cow hit the end of the rope, it would be removed surgically clean.

Hair or "mecarty" ropes were also used for hackamore ropes. Hair ropes became a highly developed art in the Intermountain area. Reins and head stalls became a specialty, and white and black blends to provide a salt and pepper effect were common. This Spanish heritage art form was practiced by Indian craftsmen, as well as prisoners at the Nevada and Idaho state prisons.

The highest state of rawhide craftsmanship was the quirt, or short whip. Hung from the saddle, it was made by braiding six to twenty-four strands of rawhide over a wooden or metal center. The top of the quirt was finished in an elaborate Turk's-head design. Confederate calvary-men rode into battle with weighted quirts as auxiliary weapons.

In each section of the Intermountain area there evolved individuals or families that became known for work with rawhide. In the Bruneau Valley of Idaho, Joe Samora, who came with the Longhorns, became well-known for eight-strand reins and sixteen-strand works of art.[44] On the North Fork of the Humboldt, the vaqueros of Pedro Altube's giant Spanish ranch included several experts in the rawhide art form.

For the professional cowboy, the danger of a breaking rawhide lariat was no laughing matter. Louis Harrell, Jasper's nephew, was roping wild horses in the O'Neil Basin in 1896 when a rawhide rope broke and hit him in the eye, causing him to lose the sight of the eye. He was roping wild horses to roach their manes and tails and sell the hair to prisoners at the state prison.[45] Longhorns and rawhide technology were introduced in the late 1860s, but rawhide technology persisted well into the twentieth century.

A ranch headquarters of the 1880s was just a point of departure. Livestock operations were mobile functions done on the open range, and there were two roundups each year in spring and fall. The spring roundup was primarily concerned with branding, marking bull calves, and earmarking slick-ear calves. The fall roundup picked up slick-ears missed by the spring passage, and sorted out the fat beef that would be marketed.

The roundups were centered around wagons. Usually there were about twenty cowboys to a wagon. Each man had five to ten horses, some broken, some green. Each cowboy was expected to have the green horses broken by the end of roundup. Some outfits hired a special broncobuster who worked on the green horses.

The wagon, driven by the cook, formed a mobile headquarters for each roundup. It would move to a given place where water and horse feed were available. The cowboys gathered cattle and worked them to a given holding area, usually a large, level flat. The cowboys ate and slept at the wagon until all cattle were worked in that area. Then, bedrolls and cooking gear were loaded for the move to the next location. The cowboys slept in bedrolls made of several blankets or quilts rolled in a canvas tarp and tied with rope. They were bulky, twice the length and two or three times the diameter of today's rolled sleeping bags.

The horse wrangler was responsible for the extra horses, and had to be up well before the regular crew to bring the horses into camp. Usually a mare was fitted with a bell to help the wrangler locate horses in the dark. A wrangler's axiom was "hear the bell and sleep well."

Sparks-Tinnin usually operated two separate roundup wagons. One worked the Thousand Spring Valley ranches in Nevada, and the other worked the Snake River side of their range. Representatives (reps) of neighboring large ranges also went along and were responsible for

111

identifying and branding calves that belonged to cows branded with their company's brand. Likewise, Sparks-Tinnin reps traveled with the Bradley and Russell wagons and other large roundups.

Once segregated by ownership, branding could begin. Calves were roped and dragged to the branding fire. On some roundups it was customary to rope by the hind feet rather than the neck. Calves caught in this manner were easier to throw and hold for branding, but a higher level of roping skill was required. A cowhand on the ground grabbed the calf by the flank, jerked it off its feet. While the horseback roper kept the rope taut, the cowboy grabbed the hind leg to keep the calf on the ground. The hot branding iron was applied, the sickening smell of burned hair filled the air. The appropriate earmark was made. If the calf was a male, it was castrated. Often the testicles were saved on a hot shovel blade at the branding fire, where the hungry hands could pick up a hot Rocky Mountain oyster. At some of the roundups in Elko County, huge numbers of cattle were worked. The Juniper Basin rodeo averaged 10,000. The workday began at 4:00 A.M., and sixteen-hour days were frequent.[46]

The entire roundup operation evolved from the trail outfits that drove Longhorns from Texas across the plains to the new Northwest ranges. A mess wagon cost approximately $125.00. The majority were "three to three and one-half inch" wagons, meaning the diameter of the axle where it entered the hub. The Bain, manufactured at Kenesha, Wisconsin, and the Peter Schulter from Chicago were favorite makes of wagons.[47]

The chuckbox at the rear of the wagon never received the dignity of a patent, but its use as standard equipment spread over the range. The endgate of the wagon bed was removed and a cupboard was built of strong lumber at the end of the wagon. It was two to three feet deep, and its perpendicular front was about four feet high. The rear wall of the box was hinged at the bottom so it could be swung down to form a work-table.

Inside, it was fitted with partitions, shelves, and drawers, and two doors folded snugly over the partitions to hold everything in place while the wagon was on the move. Each item had its place. The larger divisions were for the sourdough jar or keg, a partly used sack of flour, and bulky utensils. There were drawers for tin plates, cups, spoons, knives, and forks. Salt, pepper, soda, and baking powder were in tins with tight lids. Every cook reserved a drawer for purgatives and cure-alls like quinine, calomel pills, black draft, and horse liniment, the latter to be used on man or beast. Reserved for the cook's private use was a bottle of whiskey.

The major part of heavy supplies like flour, bacon, molasses, coffee beans, and canned goods were carried in the bed of the wagon. Attached beneath the chuckbox was another smaller box with a hinged door for Dutch ovens, pots, and skillets. One side of the wagon carried a water barrel, often wrapped in a wet gunnysack to keep contents cool.

A "fly" was also included in the cook's equipment. This was a canvas sheet which could be stretched at the end of the chuckwagon to make shade and shelter for the cook. A necessary implement for the cook was his "gouch hook" or pot hook, an iron hook used to lift the heavy lids off his cooking utensils. There was also a supply of fire hooks, or iron rods, from which cooking vessels were hung over the fire.

Camp life on roundups was based on Dutch-oven cooking. Either a rack supported the oven over the fire, or it was buried in the coals. A sourdough keg kept in the wagon provided starter for biscuits. If the crew was large and the weather cool, a steer would be killed for fresh meat. During warm weather when fresh meat would spoil, the crew was stuck with home-cured ham, shoulder, or sowbelly. When fresh meat was killed, the first night's meal would be fried liver and onions, with Dutch-oven biscuits or pan bread.[48]

One of the most famous of all western foods was the sourdough biscuit, which allowed bread-making without the use of commercial yeast. The cowboy preferred sourdough bread to any other. The cook's most particular job in preparing for the start of a roundup was to secure the proper keg for his sourdough mixture. Most cooks would defend their sourdough kegs with their lives.

When ready to make bread, the cook poured flour into a large pan until it was two-thirds full, made a deep impression in the center, and poured in his sourdough batter. Next he added a teaspoon of soda dissolved in a little warm water, plus a small amount of salt and lard. As he stirred, the cook worked the dry flour from side to side, being careful to distribute the soda and shortening thoroughly. He then drenched his worktable with flour and kneaded the batch thoroughly.

While preparing the dough, the cook had a fire going in order to have the red-hot coals burn down for his baking. He next put a

113

generous portion of lard in the Dutch oven to melt. When the oven was placed on the coals, the lid also received coals on top so that it, too, would be thoroughly heated. The Dutch oven was the most important utensil of the cook's equipment. It was a very large, deep, thick iron skillet with three legs under the bottom and a heavy lid with upturned lip fitting the top. It was also used as a skillet.

The range cook had no use for a biscuit cutter. When the grease had melted in the Dutch oven, he merely pinched off pieces of dough somewhat smaller than an egg, and rolled them into balls between his palms. In the grease, he turned them so that all sides would receive a coating to prevent their sticking together. As he placed the biscuits in the oven, he jammed each tightly against the other. The tighter they were, the higher they would rise and the lighter they would be.

The oven was then placed near the fire for thirty minutes to allow the biscuits to rise while the cook went about other preparations for the meal. When other items were nearly ready, the cook placed the oven on the coals and covered it with hot coals. Bread was better when fewer coals were used on the bottom and more on the top; then, the crust would be brown and the center tender.

Beans were a staple for all meals, including breakfast. No range cook would start his roundup wagon without a good supply of dried beans. Most cooks preferred the brown-spotted pinto beans. The black iron pot of beans cooking, with its miniature geysers throwing up little jets of steam as it bubbled over a slow fire, would excite any cowboy's appetite.

Beans were usually cooked a minimum five hours over a slow fire. Dry salt pork was cut into pieces and dropped into the pot to give seasoning. Everything was washed down with boiled coffee in pint-sized tin cups. A three- to five-gallon pot was standard size for ten to fifteen men. The sight of a wide-bottomed, smoke-blackened coffee pot on the coals, with the brown liquid bubbling down its sides, was a picture to warm any cowhand's innards.

After the cook had the rest of the meal under way, he placed the coffee pot, two-thirds full of cold water, on the coals to boil. When water reached boiling, he dumped in the correct quantity of ground coffee. After it had boiled to strength, he dashed in a little cold water to settle the grounds.

One favorite cowboy meal was made from liver, brains, heart, tongue, marrow gut, sweetbreads, kidney, onions, salt and pepper, and

114

simmered in a huge, iron pot. For some reason this was often called "son of a bitch stew." After two or three days of reheating, all the internal organs tended to lose their individuality.

When the range cook started to prepare meat for a meal, he sliced off a sufficient number of steaks, and tenderized them by pounding with a hammer on the back of a heavy knife. The cook cut suet into small pieces and put a handful into a hot Dutch oven. When the suet had cooked down, he fished out the cracklings which were left. The slabs of steak were salted, covered with flour, dropped into the sizzling fat, and the lid put on. The steaks were cooked well-done. A special treat was to sit around the campfire at night and roast beef on a willow stick.

High on a cowboy's list of luxuries was freshly baked pie. The range cook baked his pies in the Dutch oven as he did his bread. The dough was rolled out with a beer bottle, placed in a greased pie pan, filled with stewed fruit, covered with top crust, trimmed with a knife, then scalloped around the edge with a fork to seal the top and bottom crust together. On the top crust, to allow for steam to escape, he usually cut the company's brand. Spotted dog, boiled rice with raisins added, was another favorite.

The cowboys bought their own clothes and equipment. Stetsons were favorite hats, costing from seven to thirty dollars. California spurs with twin bells on each shank were popular. They cost from five to fifty dollars, depending on the amount of silver used. "Noisy" shirts were also worn, in colors like purple, red, and orange. The cowboy slickers were canvas, covered with fish oil; cost $3.25. A horse was not really 'broke,' until the cowboy could mount while wearing his slicker. The average cowboy paid $24 for a Spanish bit, $5 for a bridle, $25 for a horse, $60 for a saddle, $10 for a saddle blanket, $2.50 to $5 for a quirt, $10 for spurs, and $10 for a hat; about $150 total for his personal equipment.[49]

President Teddy Roosevelt reflected on his experience in the American West, "we who have felt the charm of the life have exulted in its abounding vigor and its bold, restless freedom, will not only regret its passing for our own sake, but must also feel real sorrow that those who come after us are not to see, as we have seen, what is perhaps the pleasantest . . . and most exciting phase of American existence."[43]

John Sparks and John Tinnin had a great and far-flung cattle empire on the sagebrush/grasslands. It reached from Wells to Pilot Peak on the south to the Snake River on the north. They owed money on huge mortgages, and their range was overstocked and overgrazed. The market for beef was severely depressed after the hard winters of 1886 and 1887. Despite all these problems, John Sparks was positive and confident when he told H.H. Bancroft that the firm had no books; that all accounts were kept in his head.[50]

Notes

1. Dictation by John Sparks on file at Bancroft Oral History Program, Bancroft Library, University of California, Berkeley, California.

2. R.G. Lillard, *Desert Challenge*, (New York: Alfred A. Knopf, 1944). This description sounds like John Sparks, and one wonders if the author, who neither cites nor mentions an original source for the information, mixed the backgrounds of the partners.

3. C.S. Scarbrough, *Land of Good Water, Takachue Pouetsu, A Williamson County, Texas History*, (Georgetown, Texas: Williamson County Son Publishers, 1974).

4. Letter from Mrs. H.G. Scoggins to Mrs. J.B. Golden, 26 June 1979. Copy of letter furnished by Mrs. Golden.

5. *Elko Weekly Independent*, 6 November 1881.

6. *Elko Weekly Independent*, 15 June 1883.

7. J.M. Guinn, *History of the State of California and Biographical Record of the San Joaquin Valley*, (Chicago, Illinois: The Chapman Publishing Co, 1905). Many modern accounts of the Jasper Harrell sale to Sparks-Tinnin mention the formation of a million-dollar stock company to finance the purchase. There are no contemporary records of this company. Writers may have been confused by the million-dollar stock company formed by Sparks-Harrell in 1891.

8. *Elko Free Press*, 1 September 1883, and *Chicago Weekly Drovers Journal*, 6 September 1883. Also see letter to *Chicago Tribune*, 7 July 1883, quoting Beecher, in *Breeders Journal IV* (August, 1883) p. 459.

9. John Clay, *My Life On the Range*, (Chicago, Illinois: Privately Printed, 1923).

10. V.S. Truett, *On the Hoof and Horn in Nevada*, (Gehrett-Truett-Hall, Los Angeles, California, 1950).

11. C.A. Brennen, assisted by C.E. Fleming, G.H. Smith, Jr., and M.R.

Bruce, *Receipts and Costs on the Nevada Range Cattle Ranches for the Years 1928, 1929, and 1930*, (Reno, Nevada: Bull. 126, Agric. Expt. Sta., University of Nevada, 1932).

12. E.S. Osgood, *The Day of the Cattleman*, (Chicago, Illinois: The University of Chicago Press, 1929).

13. C.W. Hodgson, *Idaho Range Cattle Industry*, (Moscow, Idaho: Animal Science Department, University of Idaho, 1948).

14. Truett, "On the Hoof and Horn in Nevada."

15. R.R. Elliot, *History of Nevada*, (Lincoln, Nebraska: University of Nebraska Press, 1973).

16. V.O. Goodwin, "Lewis Rice Bradley: Pioneer Nevada Cattleman and Nevada's First Cowboy Governor," *Nevada Historical Society Quarterly* (1977) XIV:11–12. Goodwin refers to these cattle as "Texas Longhorns." There is little or no evidence that Texas longhorns were regularly reaching Missouri in 1852. These were probably native American cattle or work oxen. Taking beef cattle to California in 1852 was akin to carrying coals to Newcastle.

17. Ibid. Goodwin said Bradley moved to Nevada because of floods; however, Myron Angel, ed., in *History of Nevada*, (Berkeley, California: Howell-North, reprinted edition, 1881), says Bradley moved because of drought. The same version appears in Buster King, "History of Lander County," M.S. thesis, University of Nevada, Reno, Nevada, 1954.

18. E.F. Treadwell, *The Cattle King*, (New York: The MacMillan Co., 1931).

19. *Humboldt Register*, 15 January 1870.

20. J.M. Townley, "Reclamation in Nevada, 1850–1904," Ph.D. diss., University of Nevada, Reno, Nevada, 1976.

21. C.B. Creel, *A History of Nevada Agriculture*, (Reno, Nevada: Max C. Fleischmann College of Agriculture, University of Nevada, 1964).

22. C.W. Gordon, "Report on Cattle, Sheep and Swine," *Supplementary to Enumeration of Livestock on Farms in 1880*, 10th Census of the United States, Vol. III, (Washington, D.C.: GPO, 1880).

23. Clay, "My Life on the Range."

24. *Elko Weekly Independent*, 23 October 1887.

25. *Daily Nevada State Journal*, 15 July 1885.

26. O.B. Peake, *The Colorado Range Cattle Industry*, (Glendale, California: The Arthur H. Clark Co., 1937).

27. H. Goding, and H.J. Raub, *The 28 Hour Law Regulating Interstate Transportation of Livestock: Its Purpose, Requirements, and Enforcement*, (Washington, D.C.: GPO, Bull. No. 489, U. S. Dept. of Agric., 1918).

28. J.R. Williams, *Cowboys Out Our Way*, (New York: Charles Scribner and Sons, 1951).

29. Gordon, "Report on Cattle, Sheep and Swine."

30. Ibid.

31. Truett, "On the Hoof and Horn in Nevada."

32. E.B. Patterson, L.A. Ulph, and Victor Goodwin, *Nevada's Northeast Frontier*, (Sparks, Nevada: Western Printing and Publishing, 1969).

33. Ibid.

34. Gordon, "Report on Cattle, Sheep and Swine."

35. *Report from the Chief of the Bureau of Statistics, 1884–1885*, (Washington, D.C.: GPO, Ex. Doc. No. 267, 48th Congress, 2nd Session, House of Representatives, 1885). In response to a resolution of the House calling for information in regard to the range and ranch cattle traffic in the western states and territories. Hereafter cited as "Nimmo Report."

36. "Nimmo Report."

37. Maurice Frink, *Cow Country Cavalcade*, (Denver, Colorado: Old West Publishing Co., 1954).

38. "Gordon Report."

39. Patterson, et al., "Nevada Northeast Frontier."

40. *Carson Morning Appeal*, 4 December 1888.

41. N.S. Yost, *The Call of the Range—The Story of the Nebraska Stock Grower's Association*, (Denver, Colorado: Sage Books, 1966).

42. George Stewart, "History of Range Use," pp. 119–233, in *The Western Range*, Washington, D.C.: GPO, Senate Doc. No. 199, 74th Congress, 2nd Session, 1936).

43. W.M. Raine and W.C. Barnes, *Cattle*, (New York: Grosset and Dunlap, 1930).

44. Adelaide Hawes, *The Valley of Tall Grass*, (Bruneau, Idaho: Private Printing, 1950).

45. Personal communication to James A. Young from Newton J. Harrell, the son of Louis Harrell, Twin Falls, Idaho, July 1979. After the accident, Louis Harrell rode by himself for two days to Montello, Nevada, where he boarded the train for Ogden, Utah. The hospital at Ogden succeeded in saving the sight

of his other eye. While recovering in Ogden, Louis Harrell walked down to Browning Brothers Gunsmiths and purchased a new Colt single-action revolver. The revolver, original box, and receipt, are the possessions of Newton Harrell.

46. Peake, "The Colorado Range Cattle Industry."

47. R.F. Adams, *Come and Get It*, (Norman, Oklahoma: University of Oklahoma Press, 1952).

48. E.E. Dale, "Cowboy Cookery," *American Hereford Journal*, (1946) XXXVI(17): 37–38.

49. Peake, "The Colorado Range Cattle Industry."

50. John Sparks's dictation to Bancroft. On file at Bancroft Library University of California, Berkeley, California.

13. These are typical land forms of the central Great Basin with a desert flat in the foreground, an alluvial fan at the base of the escarpment, and the towering Simpson Park Range in the background.

14. In a land without forests, the early ranchers constructed houses from native stone.

15. Native stone construction was a labor-intensive process.

16. Aspen or juniper logs were used for rafters in early roof construction. Woven willow and Great Basin wildrye were used for thatch under a layer of dirt.

17. Early settlers used materials that were at hand such as these pinyon/juniper trunks used as palisades for stock corrals.

18. Saddle horses were a vital part of range livestock operations. These Sparks-Harrell horses were corralled on Golliher Mountain.

19. Because there was no refrigeration, slaughtered beef was often shared with a neighboring ranch. Note the liver hung on the post.

20. Elko, Nevada, became the commercial center for the developing livestock industry after the completion of the Central Pacific Railroad along the Humboldt River route.

21. After the hard winter of 1889-90, ranch sheep increased in importance on the sagebrush/grasslands. Competition between sheepmen and established ranchers became severe.

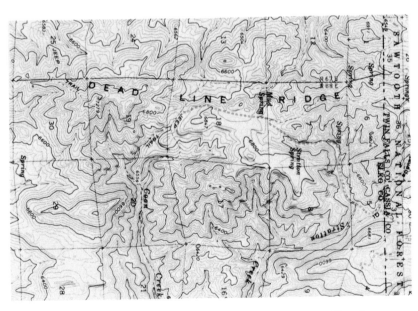

22. John Sparks established a deadline in Goose Creek Basin and hired gunmen to enforce the rule that sheepmen were not welcome beyond this point.

23. The rangelands were the real losers in the sheep versus cattle wars. This high-elevation range was practically denuded.

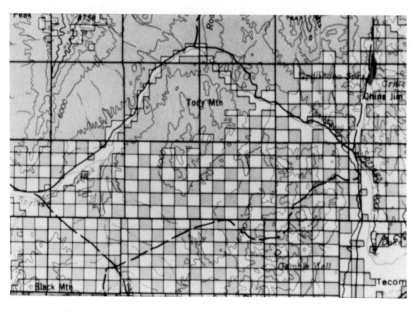

24. The acquiring of land and water was paramount. Sparks bought ranches in the Thousand Springs Creek area along the margin of the checkerboard land grant given to the Central Pacific Railroad.

CHAPTER
7

WHITE WINTER: WHITE HILLS OF BONES

Rarely does one climatological event alter plant and animal ecology, or change the social and economic structure of a wide geographical area. However, such a far-reaching and dynamic event was the devastating winter of 1889–90 in the sagebrush/grasslands of western North America.

Herdsmen are traditionally resistant to change, and they thrive on the repetitive cycles of new grass, calving, branding, marking, marketing. Traditional systems of management often persist long past their time in defiance of economic laws. Although the passing of generations is required to bring changes in established procedures, this was not the case with open-range management following the winter of 1889–90. It brought a white wave of disaster and hardship.

In hindsight, it is easy to suggest that late nineteenth-century ranchers had seen evidence that open ranging of livestock without conservation of forage for winter feeding was to invite disaster. However, at the time, the open range appeared to be an exciting, romantic, and economically attractive industry. Technically, Anglo-Texas pioneer ranchers were about to exploit the grazing resources of a semiarid environment, using techniques evolved by Spanish herdsmen in a Mediterranean environment.

Frank Dobie devoted a volume to the anatomy, culture, and color of the Longhorns. In a footnote he offers the ecological classic that the millions of Longhorns accumulated in Texas before and during the Civil War were not "free roaming animals of the plains." Completing their life cycles without the aid of man, they were creatures of oak woodlands, brushlands, and the woods of southern and eastern Texas. These areas were more favorable winter environments than the open plains. When the Texas longhorns were driven to the open grasslands of the plains, and prevented from drifting south with winter storms, this altered their free-roaming adaptation and paved the way for eventual disaster.[1]

If the open-range system the Anglo-Texans borrowed from the Spaniards had been moved gradually north until excessive winter losses were encountered, winter disasters could be attributed to experimentation, and observation of the results in a learning process. This was not the case. Ranchers continued to push north and west despite serious losses, such as cattle suffering and dying in quantities as they wintered on prairies around Kansas railheads. Ranchers were slow to appreciate or accept the risks involved in open wintering. Factors included year-to-year variance in severity, regional differences any given winter, and the changing relationships between range condition, amount, and quality of forage.

The winter of 1885–86 was severe on the southern Great Plains, but relatively mild in Wyoming and the new northwest range states. Winter losses in Kansas, Indian Territory, and the Texas plains were commonplace. George B. Louis, in written testimony for the Nimmo Report, reported losses in some parts of Texas for 1885 and 1886 as approaching 30 to 40 percent. In August of 1885 over 200,000 head were forced out of Indian Territory, by an edict of President Cleveland, and moved to already heavily grazed ranges of adjoining states. They were about to face one of the most severe winters in history.[2]

A factor contributive to the large winterkill on the southern Great Plains in 1885–86 was the extensive fencing of range with barbed wire. Patented in 1874, barbed wire was quickly accepted and by 1880 annual production reached 40,000 tons. The fences in northern Texas were designed to curb the southward movement of range cattle before winter storms. Longhorns differed from American bison in a basic instinct that had great influence on winter survival. The bison faced or drifted into the storms, but the Longhorn turned tail and drifted south. In the winter of 1885–86, barbed-wire fences hindered the southward drift of the Longhorns that piled up and died by thousands.[3]

After the 1850s, when it was discovered that cattle could winter on the high plains of Wyoming, there was an escalation of winterkills. This was proceeded by increased numbers of livestock, and declining range conditions.[4] John Clay clearly recognized that higher death losses resulted from wintering livestock on poor range, but this did not prevent him from joining the rush to disaster. Why did herdsmen whose entire heritage and training called for the maintenance of animals in fenced areas, or under constant herd supervision, suddenly turn their thousands of animals loose in the open range? By contrast, the Mormon settlers' first agricultural communities established in Utah in the 1850s followed the northern European practice of communal herding during the day and the return to village by night.[5]

The open ranging of livestock on the plains, as practiced by the Spanish in the New World, was sold to the new wave of ranchers, and one of the more enthusiastic salesmen was Dr. Hiram Latham, a medical doctor for the Union Pacific Railroad stationed at Laramie, Wyoming. In the 1860s, Dr. Latham wrote a series of letters to the *Daily Herald* of Omaha, Nebraska, describing the Laramie plains and their potential for livestock. A pamphlet was then widely distributed by the Union Pacific Railroad to attract settlers. Dr. Latham describes the Laramie plains, "The grasses are self-curing, and sheep and cattle live and thrive year round without other food or shelter than that afforded by nature." He cited areas of supposedly similar climate afforded in South America and South Africa where cattle were open ranged, then reinforced his arguments with quotations from the Bible. Dr. Latham goes on, "The secret of these great herds of cattle, horses, and sheep for so many centuries is *winter grazing.* I speak of the grazing in all these countries to show that the idea of cattle grazing in winter in the latitude and altitude of these plains is not new, but as old as the history of man."[6]

The winter of 1886–87 again denied Dr. Latham's prophecy. Total rainfall for May, June, and July 1886 was 2.55 inches at Cheyenne compared to 5.15 for these three months during the twelve summers from 1875 to 1886. The number of calves counted had been very low at the spring roundup in 1886 in Wyoming, as late spring storms had decimated the calf crop. Montana ranchers reported death losses attributed to cattle grazing poisonous plants. Granville Stuart blamed losses on the drought and overgrazing of the desirable forage species.[7]

Wildfires were commonplace in Wyoming during the summer of 1886. Fire consumed grass that would be vitally needed during the coming winter. Big fires occurred along the foot of the Judith range and on the Musselshell in Montana, and the sky was often obscured by smoke and dust. John Clay considered the summer of 1886 the driest he had experienced in thirty-five years of livestock activity in Wyoming. Streams as large as the Rosebud ceased to flow. Because of drought and overstocking, range animals approached the winter in poor condition. Many were recent arrivals from Texas, new to the ranges on which they were to be wintered.

Writers concurred that there was overgrazing of the range during the summer of 1886. John Clay rode over the south central Wyoming range and saw "scarce a blade of grass." The same conditions prevailed, he said, on the Belle Fouche, Little Missouri, and Powder rivers. Additionally, assessment figures indicate that there were three times as many cattle in Wyoming in 1866 as there were to be eight years later.

Between January 28 and January 30, 1887, a record blizzard swept down from the north. Eastern Montana ranchers felt the Arctic had suddenly enveloped them. It was -46°F. Cattle drifted before the storm. Even fat steers froze to death along the trails. Inhabitants of Great Falls, Montana, looked out through the swirling snow one morning to see the gaunt and reeling leaders of a herd of 5,000 cattle that had drifted south from the frozen Missouri River. Cattle drifted through the streets, standing about the livery stable where they could pick up a few wisps of straw.[8]

Weather Bureau records at Bismark, North Dakota, show that mean temperatures for January and February 1887 were 12.4° and 12.9°F below normal.[9] In log ranch houses and sidehill dugouts ranchers and cowboys tried to block out the bawling of hungry cattle, bunched at the corral fences crying for hay that was not available. Ranch employees were found frozen to death near Sundance, Evanston, and Stinking Water, Wyoming. Longing for another warm "chinook wind" became the yearning for a miracle.

Charles Russell and Jesse Phelps were looking after 5,000 head of Kaufman and Stadler cattle. Louis Kaufman wrote Phelps a letter to request information on how the cattle were surviving. This inspired the famous Charles Russell drawing "Waiting for the Chinook." In the drawing, a starved-looking steer stands humped over in the snow, barely able to stand, while hungry coyotes await the meal soon to be theirs. Russell and Phelps sent the drawing to their bosses without explanation. When it was received in Helena, it caused considerable excitement. Someone added the subtitle, "The last of 5,000." This drawing became the symbol of the decline of open ranging of livestock on northern plains.[10]

Below-zero temperatures and wind continued well into the spring season, as cattle cut their legs on the crusted snow and the scarcity of water contributed to winter losses. Seeking water in the air holes of the Yellowstone River, cattle were crowded forward into icy water by the pressure of others. The small creeks and springs of the ranges were frozen. When spring rains did arrive, they were very heavy, and the weak animals were prey to miring in the mud along the streams.[11]

Old-timers, hardened to losses on range operations, came near to panic in the spring of 1887. Even many of the bright young men from the halls of ivy, or the drawing rooms of England, were shaken by the carnage on the range. A fascinating business had suddenly become distasteful, and many gave up or pulled out.

Extent of losses in the winter of 1886–87 was difficult to verify. Some counts were greatly inflated to investors. Some ranchers lost nearly all their herds, especially if they were recent arrivals. Overall, the Wyoming losses may not have exceeded 15 percent. But the winter caused a loss of confidence by creditors, and liquidation of starving herds at ruinously low prices.[12]

Estimate of Montana losses went as high as $5,000,000. To meet them, every available steer was shipped. Practically none were in condition for immediate slaughter and fodder was in short supply in the Corn Belt. There was little demand for feeders. Chicago prices took another drop.

John Clay suffered a 25 to 30 percent death loss on the "moccasin" steers he had bought for Sparks and Tinnin for the 71 Ranch, before he suffered the additional blow of selling the survivors for less than he had paid. Cattle worth an average $9.35 per hundred weight on the Chicago market in 1882 brought $1.00 per hundred weight in 1887.[13]

Newspaper editors in the spring of 1887 castigated the giant ranching companies. "Insatiable greed" was a phrase used to explain their view of the winter disaster. Companies financed with foreign capital were especially subject to abuse in the press.

In his published reminiscences, Granville Stuart was bitter about the treatment he received from the newspapers. He disclosed that while many large ranchers in eastern Montana were losing their own herds, they nevertheless supplied hay to homesteads so the family milk cow could survive the winter. The farmer-ranchers who had a few cows and sufficient hay benefited by the misfortunes of the large ranchers.

The range in 1887 was often nearly devoid of stock from the combination of winterkill and forced sales. The vegetation dynamically responded to the abundant precipitation, and probably to the reduced density of plants caused by drought mortality. The greatly reduced cattle population of 1887 also brought about a surplus of cowboys. Unemployment produces many evils, and the wild, young cowhands who had been viewed as "knights of the plains" during the previous decade became suspect when calf counts at roundups began to drop alarmingly. Unemployed cowboys-turned-homesteaders sometimes viewed unbranded calves of their former employers as free game.

Much has been written about the economic and social impact of the drought of 1886 and the hard winter of 1886–87 on the expanding range livestock industry. But the impact on the basic resource of plant communities and supporting soils seems to have been overlooked. John Clay, recognized at the time as an outstanding leader of the industry, spoke of tighter credit as the key to solving the industry's ills. The idea of range management did not even surface. Tight money slowed expansion and may have reduced stocking rates, but it was like giving aspirin to a man with a broken leg.

West of the Rockies, the winter of 1886–87 was rather open and mild. Did Intermountain ranchers learn from the disasters east of the Rockies? John Sparks had ranching interests in Wyoming, so there is no question he was aware of the danger of overextending livestock on the open range without adequate hay reserves.

One of the reasons Sparks moved his operations to Nevada was to take advantage of the Great Basin topography. There, altitudes of some of the valleys were sufficiently low to support desert winter ranges, in contrast to the higher altitude plains of eastern Wyoming. Also, the more arid climate of the valleys appeared to assure safety from winter disasters. Jasper Harrell had let cattle range freely in that area for a decade with an estimated winter loss of 1 percent! Again, with hindsight, Sparks and Tinnin might have considered more carefully the history of the previous two decades.

During the disturbances in Utah from 1856 to 1858, large numbers of federal troops were stationed in the territory. The freighting firm of Russell, Majors, and Waddell contracted with the United States government to supply beef. Alexander Majors tried to winter 3,500 steers in the Ruby Valley of Nevada during the winter of 1859–60. November brought a heavy snow with cold temperatures. In forty days, all but two hundred steers lay in starved and frozen heaps. The next hard winter was 1861–62. Of a herd of 3,000 ewes wintered on the Truckee meadows, only 500 survived after over two feet of crusted snow clogged the valley, and attempts to break trail to the Pyramid Lake desert failed.[14] A colloquial comment in the Great Basin is that "only fools and tenderfeet predict the weather." A grain of truth is carried in this message due to the great variance in the timing and quantity of precipitation received.

Based on tree-ring growth and existing records, rainfall maxima occurred in 1853, 1862, 1864, 1868, and 1890 through 1893 in the Great Basin. Minima occurred during the 1840s, 1864, 1871, 1889, and 1898. Maxima or hard winters came six to seventeen years apart, with most of the intervals at six, seven, or eight years. Intervals between minima or dry years were most frequently six to seven years apart or multiples of these. There was, usually, a swift succession of dry and wet times.[15] Essentially, average climatic conditions are the result of contrasting extremes. The Sierra Nevada blocks winter storm fronts from the Pacific Ocean, and cast a rain shadow across much of the Great Basin. Weather on the west slope of the Sierra Nevada also had important impact on development of the early ranches in Nevada.

Severe drought in California was a factor favoring the rapid stocking of Nevada ranges. The green-feed period for the annual ranges of cismontane California, with its Mediterranean-type climate, occurs from October to May. Cattle are dependent on dry feed during the summer drought. If the winter is dry and there is no accumulation of dry feed to carry the stock through the summer, disaster will occur unless other sources of forage are obtained. During the late nineteenth century the feed was secured by driving the cattle to the virgin sagebrush/grasslands of the Great Basin. Dry winters in California usually mean below-average growing season moisture in Nevada. When Nevada ranges were pristine, the consequences of this livestock transfer became the subtle degrading of plant communities. As Nevada ranges were exhausted, such transfers were disastrous for both the livestock and plant communities.

In Idaho and northern Nevada, 1879–80 brought the most severe winter after 1864. Estimates of losses ranged from 6 percent for Nevada as a whole to 20 percent for northern Nevada or south central Idaho. Based on tax rolls, the decline in cattle herds from 1879 to 1880 was 50 percent. Tax rolls are, however, a very poor source for nineteenth-century livestock numbers. There was, however, an exodus of ranchers from the Great Basin in 1880.[16]

Humboldt Valley ranchers attempted to winter cattle on the Owyhee Desert during 1879–80. There was little snow on the desert floor during most of the winter. Cattle stayed close to existing water

127

supplies because of the lack of water in ephemeral streams. The forage, mainly winterfat or Indian ricegrass, was entirely consumed in the vicinity of the watering points. Extreme cold contributed to losses in the starving cattle. The irony of the situation was that while the cattle starved, there was standing forage on the alluvial fans located excessive distances from stock water.[17]

There is evidence that Sparks-Tinnin realized they were nearing or exceeding grazing capacity of the range during the late 1880s. Newspapers reported the northeastern Nevada giants of the livestock industry were trying to reduce livestock numbers and upgrade the quality of their stock.[18] The cattle industry in northeastern Nevada had enjoyed good years from 1880 to 1886, and earlier hard winters were forgotten. From 1886 to 1889, precipitation was below normal. Stocking rates exceeded forage supplies during the growing season and ranchers including Sparks and Tinnin found it necessary to ship cattle before they were ready for market.[19]

In March of 1889, Starr Valley, Elko County, ranchers reported ranges in good condition, but by the end of May reports in the Elko newspaper expressed concern. The growing season for herbaceous vegetation in the northern Great Basin extends from early spring when temperatures warm sufficiently for plant growth until soil moisture is exhausted in early summer. In the spring of 1889, soil moisture was exhausted in early May on lower elevation sagebrush ranges. Perennial grasses withered and lapsed into dormancy. Streams that in the memories of the oldest settlers had been perennial, shrank to interrupted pools, then dried completely.

Perennial snowbanks that occupied the glacial cirques on the towering mountain masses that form the headwaters of the Humboldt River shrank and disappeared. For the first time in memory there was no visible snow on the Ruby Mountains. Hulking like a naked giant in the eastern skyline, the Ruby Mountains provided a reminder to the residents of Elko that they were participants in an environmental event of unusual occurrence.

Precipitation for the winter (1888–89) was reported in the Elko papers as 3.5 inches. (Mean annual precipitation, July 1 through June 30, for the Elko period 1870 through 1915 was 9.09 inches.) The weather records for Elko indicate 6.35 inches; however, this included a large local thunderstorm of 2.80 inches occurring in late May after the herbaceous range plants had withered and dried.[20] The summer months of 1889 were exceptionally hot, evidenced by dust, withered vegetation, and dried streams. The city of Elko was faced with a severe water shortage. The free-roaming horses were concentrated by reduced water places, and movement of bands could be traced for miles by dust clouds.

Autumn and Indian summer is the glorious period in the Great Basin. Cloudless warm days alternate with crisp nights. Aspen leaves form golden cascades in the draws on the high mountains. The gallery forests along the Truckee, Carson, and Walker rivers form golden arches over still pools. On October 13, 1889, the first rains broke the drought and settled the dust. Through November the weather was ideal, until cold rains changed to wet snow.

On December 5, the white winter struck northern Nevada with full fury, with blizzard conditions for seven consecutive days. The editor of the Elko paper welcomed the rains, "Rainfall this fall is equal to that of all last winter. One foot or more of snow in the mountains last night." There was excitement throughout Nevada, after the continuing drought of the previous three years.[21]

Feed supplies, however, were exhausted by the holidays. Just before Christmas, 1889, the season's first snowplow passed through Elko clearing the Central Pacific tracks of six inches of snow. The paper welcomed the "assurance of a prosperous New Year." But after Christmas, the paper began to express concern. Valleys north and south of Wells were "belly deep to a horse." Just before the New Year, eighteen inches fell in Elko on Monday and continued through Wednesday. For the first time since 1862, when Ruby Valley, Elko County, was settled, residents did not receive mail deliveries.[22] By early January 1890 snow was two feet deep in the valleys and crusted. January 6, 1890, saw -40°F in Elko, warming six days later to -36°F. Houses creaked in the night as contracting timbers pulled square-cut nails. Snow north of Elko was reported to be forty-two inches deep, and the stage to Tuscarora was stranded.

The stage to Twin Falls was the communications link for the Sparks and Tinnin ranches. The Salmon Falls Creek area around San Jacinto, a major Sparks and Tinnin ranch, is recorded as the coldest part of Nevada, with a -50°F recorded and average annual snowfall of 28.6 inches.

The January 12 edition of the Elko paper reported ranchers were

129

planning to ship their cattle to California, but this received a setback when the tracks were blocked both east and west to Elko. That same week, the Elko editor found a half-inch of ice on the water in his well.[23]

In mid-January, northern Nevada received six more inches of snow. Halleck Cattle Company cowboys wore all the clothes they owned, pushed their starving cattle through frozen willows along the Humboldt River to isolated patches of feed. Sloughs piled up with dead cattle. At a Humboldt bridge, thirty-nine cattle were caught at one time. On January 15, the Elko paper reported several thermometers had registered near -60°F. The official reading was -42°F.

Low temperatures continued through February, at -40°F on the first day, with -41°F recorded on the last day of the month, with additional new snow. Nevada Land and Cattle Company estimated their winter losses had already reached 98 percent, and hay they had stored could not be moved over the drifted roads. In Secret Valley, Elko County, A.G. Dawley reported fifteen-foot drifts between his house and barn, where he carried feed and water through a tunnel to save two stallions.

There is a love-hate relationship between herdsmen and their domesticated charges. They take pride in their animals from birth to maturity, then ship them off to slaughter. Only the most calloused rancher could stand by while his animals suffered without becoming emotionally involved. The small homesteader with twenty cows knew each animal well. Families had scraped, saved, and done without to accumulate a few cattle. Also, they represented the means of paying off supplies bought on credit, that rare new dress for overworked wives, and any toys for next Christmas.

On those -40°F, -50°F, or -60°F nights, homesteaders herded their animals around bonfires and fed handfuls of native-grass hay, cut the previous summer by hand, to the weakest animals. In an area where structural timbers and lumber had to be shipped great distances, houses, sheds, and outbuildings were built from native stone. These buildings were roofed with juniper or aspen rafters, thatched with woven willow sticks and Great Basin wildrye stems, and covered with dirt. In the desperate white winter of 1889–90, ranchers ripped down the roofs to salvage feed to keep a saddle horse alive a few more days.

By late January, things were very bleak throughout the Intermountain area. Elko townspeople shoveled snow from roofs and canceled church. Finally, in desperation, cattle turned to browsing the sagebrush, which inhibits the activity of rumen microorganisms. Forced to make sagebrush a large percentage of their diet, they die of the malady Nevada ranchers call "hollow belly." An autopsy reveals a rumen packed with sagebrush rich in nutrients, but indigestible. The rumen is a vital life-sustaining factor in their winter survival. Action of the rumen microflora, in breaking down high-cellulose-content forage, creates heat. A cow with a full rumen can withstand bitter cold. An empty rumen means a cold cow, soon dead.

The question was no longer whether many cattle would die, only how. One herd of 300 broke into a stockyard, and 117 smothered in the crush to reach the hay. Twenty-six cows crowded into a Starr Valley cave to escape the storm; all perished. Horses bunched up to chew each other's manes and tails until all the hair was gone; then, died in a group. Animals marched up and down, seeking the herdsmen to which their ancestors, millenniums ago, had given bondage in return for care and subsistence. The herdsmen were not able to fulfill their responsibility, and animals died by the thousands.

Late in January, L.A. Nelson, in charge of Sparks and Tinnin cattle on the Salmon River, reported in the Elko paper, "We have a foot of snow on the winter range. Cattle have been shrinking very much the last 10 days. Unless we have a thaw in a short time, there will be a good many of the old cows that will turn up their toes before spring, as there is no hay to feed anything but saddle horses." On the North Fork of the Humboldt and in Independence Valley, even sheltered cattle on feed were reported dying from the cold.[24]

In mid-February, the editor of the Elko paper interviewed Sparks as he passed through Elko, returning from an inspection of his ranches. Sparks cautioned the editor to keep a stiff upper lip and to avoid printing scare stories in the paper. He said his firm had lost more stock on two previous occasions than they had that winter. He had "traveled to the Snake River and back looking for dead cattle and found few except for thin and weak stock." He said they were feeding 18,000 head.[25] This is surprising, as one of his foremen reported they were out of hay in late January. The editor concluded that Mr. Sparks was not guessing, had seen the range, and understood range country. Sparks believed that ranchers who shipped their cattle out in the fall would lose more money than if they had left them on the range and shipped hay in. He repeated his story when he reached Reno, saying "he had ridden 400 miles on horseback without seeing many dead cattle."[25]

On February 10 the same paper had carried a story reporting that N.H.A. Mason had spent two months in the snow in a futile attempt to save his herds. The winter broke Mason, and resulted in his interests being absorbed by Miller and Lux.

Rail communications suffered. February 9 saw the first train reach Elko from San Francisco since January 15, when snowsheds on the Sierra Nevada were swept away by snowslides. The city of Elko used rail cars to haul snow out of town.

On March 2, 1890, several thermometers registered -40°F, just ahead of the moderate temperatures that turned the deep snow to slush. The 17th of March is a day that stands out even in that formidable winter. Storms over southern Idaho and northern Nevada began with rain and sleet, then turned to snow.[27] Drifts of heavy slush on the lower sagebrush ranges refroze with icy-hard crusts. Rain and sleet saturated the shaggy winter coats of livestock that had survived, and were searching for forage on the open range. The temperature dropped and the weakened animals were unable to shake the ice from their coats.[27] The Nimmo Report listed late spring storms as a separate category of death losses for the range livestock industry.

The white winter of 1889–90 was severe in virtually all states and territories west of the Rockies. The Pacific Northwest had one of the four worst winters recorded before 1900, and the Governor of Idaho labeled it, "the most severe winter ever experienced since settlement of this country."

Similar reports came from the Big Bend of the Columbia, to the Lake County in south-central Oregon, and from the Yakima Valley to the Snake River Plains. The same picture came from the Palouse country southward through the Walla Walla Valley and northeastern Oregon to the Owyhee and Malheur counties. It was a gruesome pattern of cattle dying from lack of feed, water, and shelter.[28] The valleys of central California became lakes during this classic winter of 1889–90. Henry Miller spent the entire winter rushing from the stemming of one disaster to cope with the next.[29]

The L-7 Ranch near Baggs, in western Wyoming, was owned by one of the Swan brothers, a family closely associated with the winter disaters of 1886 and 1887. L-7 cowboys drove their cattle onto the Red Desert north of Rawlings to let them rustle for themselves on the desert shrubs. On the day before Christmas, 1889, they turned loose 10,000 cows. Their losses were estimated at 75 percent of the cattle and 66 percent of the saddle horses.[30] One newspaper editor wrote that one consolation for the ranchers of the Far West was "they all went down together."

The regional extent of the disastrous winter of 1889–90 made it difficult to restock the ranges in 1890. The trail drives from Texas had been shut down by the settlement of western Kansas and the Texas fever controversy. Nature's horrors fade with time, and in April, 1890, spring came to the sagebrush/grasslands. Riders covered their noses with bandannas to stifle the stench of carcasses thawing in the retreating snowdrifts.[31] East of Elko, one drift revealed five cows, one horse, two mule deer, and one pronghorn huddled together in death.

It was said that a man could walk on dead cattle for 100 miles along the Marys River fork of the Humboldt. J.D. Bradley of Mason and Bradley reported the country around Deeth and Halleck was strewn with dead cattle. He told the Elko paper the estimated loss of $750,000 in stock for the country was entirely too low. J.R. Bradley reported counting 400 carcasses of dead cows within 400 yards.[32] A.G. Dawley from Halleck expressed fears that the winter would end the cattle business in Ruby Valley.[33] A rancher wrote from Ruby Valley to the Elko papers, "I've just come in from four days riding—counted 100 dead cattle along the roads." Horses that survived had hair and skin worn off legs and noses from rooting for feed.

Carrion feeders waxed fat and happy with their choice of thousands of carcasses. Cattle and horse carcasses were found wedged in the top of juniper trees where they had walked out on the top of snowdrifts. When the ice thawed, the streams were choked with carcasses. In the Humboldt River, cattle jammed against the bridges. Downstream towns complained of the stench. People who depended on the streams for domestic water had to quickly sink wells to save themselves from a greater epidemic.

In the spring of 1890, a district judge went to hold court at Challis, Idaho. He found the stench through part of Lost River Valley so bad, he issued court order to county officers to get the carcasses burned and buried.

Estimates published in the Elko papers reported some large ranches had lost 95 percent of their livestock. One of the often-quoted statistics is that Sparks and Tinnin branded 38,000 calves during the 1885 roundup on their Nevada and Idaho holdings. In 1890 on the same ranges, they branded 68 calves.[34] John Sparks later described his losses to a reporter for *Harper's Weekly*. "In 1889 and 1890, Mr. John Tinnin and myself were ranging in Elko County, Nevada, and Cassia County, Idaho, 65,000 head of cattle—we lost that winter, which was a severe one, 35,000 head of cattle." Sparks made this comment in 1902, or twelve years after the white winter event.[35]

After spring roundup in 1890, John Sparks, Jasper Harrell, Andrew J. Harrell, and John Tinnin met in a cabin on Cottonwood Creek south of Twin Falls, Idaho. John Tinnin, the colonel of the Confederacy with the steamboat Gothic mansion in Georgetown, Texas, was broke. Jasper Harrell had sufficient capital and confidence to continue in the livestock business. Sparks must have had adequate resources as well, because he handed Tinnin $45,000 cash for his interests, and suggested Tinnin go to the sandhills of Nebraska to start again. He also promised to send Tinnin cattle to stock his range.[36]

A new company was then incorporated in California, as the Sparks-Harrell Company, with a capital stock value of $1,000,000. According to the incorporation papers, stock was fully subscribed with John Sparks subscribing to 50 percent, Jasper Harrell 49 percent, Andrew Harrell 0.8 percent, Martha and Ella Harrell 0.1 percent each. The Harrells listed their home address as Visalia, California; Sparks as Georgetown, Williamson County, Texas.[37]

The transplanted Spanish system of open-range livestock was dead. Many of the stockmen who had brought Spanish longhorns from California or Texas to the sagebrush/grasslands were wiped out. The Spanish vaquero from Texas or California, and the cowboys from Texas drifted away in the wake of the entrepreneurs who had introduced their culture to the sagebrush/grasslands. The Spanish left a sprinkling of place-names, a rich vocabulary of technical terms concerning horses, riding equipment, and the handling of livestock on the open range. Most importantly, the vaquero imparted basic skills necessary to work with cattle and horses on the open range.

The winter of 1889–90 made it apparent that conservation of forage for wintering livestock was a necessity in the sagebrush/grasslands. This required cultured hay, stored for winter. Forage crops for hay production in the sagebrush/grasslands won't grow without irrigation. Only 2 to 5 percent of northern Nevada is irrigable with water diverted from streamflow. The tiny portion of landscape that could be irrigated became the controlling element for vast areas of rangeland.

The loss of cattle and subsequent vacant ranges had an influence on the livestock industry. Critical conflicts surfaced. The range sheep industry had fewer losses because sheep were wintered exclusively on desert ranges where they were better adapted to the environment and forage base. The range sheep industry was also smaller, with fewer marginal operations. After 1900, when the range sheep industry peaked, there were many examples of excessive winter losses. But the net immediate effect of the white winter was freedom for the range sheep industry to expand without competition from previously established cattle ranches.[38]

The superabundant precipitation of the hard white winter also conditioned excellent plant growth during 1890. The ranges were virtually empty. The pristine plant communities of the sagebrush/grasslands had been severely and selectively reduced by two decades of unlimited livestock grazing. The herbaceous portion of the plant communities, especially the perennial grasses, had been selectively exploited by domestic livestock. This left the shrubs to take advantage of the near vacuum in the spring of 1890. Shrub establishment included stands of the desirable browse species bitterbrush, but it also included an overwhelming abundance of big sagebrush. Its special protection from browsing by the essential oil content of its herbage brought a basic change in the forage resources of the sagebrush/grasslands.

Pinyon and juniper woodlands greatly expanded their ranges during the decade of the 1890s. Many of the woodlands in Nevada had been severely exploited during the 1870s and 1880s as energy sources for the mining industry.[39] The combination of three good precipitation years at the start of the decade also favored woody vegetation.

Grasslands of the high plains responded dynamically after the drought of 1886 and the abundant precipitation of the winter of 1886–87. Many of the dominant grasses of the high plains grasslands are rhizomatous and responded vegetatively to occupy the environment potential released by drought losses once the rains returned. In the Great Basin, perennial grasses are largely bunchgrass and depend on

135

seed for reestablishment. Two decades of excessive grazing had virtually eliminated seed production in many areas. With seed reserves in the soil exhausted, there was no way for the grasses to respond to the abundant precipitation of the winter of 1889–90. After the reduction in livestock numbers caused by the winterkill, the perennial grasses doubtless produced abundant seed crops in 1890, but shrubs, pinyons, and junipers had the advantage of earlier establishment and preempted the released environmental potential.

The great stillness on the ranges in the spring of 1890 was broken during the summer and fall by the sounds of the pack mules and wagons of bone pickers, a profession born in the wake of the near extinction of the American bison on the Great Plains. They built white hills of bones along the railroad sidings at Montello, Toana, and Wells, following the white winter. Remains of the great expectations of the early livestock men of the sagebrush/grasslands were boiled for oil, fertilizer, and cut into buttons by Pacific Fertilizer company on the shores of San Francisco Bay.

Notes

1. F.J. Dobie, *The Longhorn*, (Boston, Massachusetts: Little Brown & Co., 1941). Although primarily known for colorful descriptions of Longhorns, this volume contains valuable ecological material in the bibliographical notes for chapters.

2. "Nimmo Report," Executive Document No. 267, 48th Congress, 2nd Session, House of Representatives, (Washington, D.C.: GPO, 1885). Prepared by Joseph Nimmo, Jr., Chief of the U. S. Bureau of Statistics.

3. Patents for barbed wire and machines for making the wire were granted to J.F. Clidden of De Kalb, Illinois, in 1874. See E.S. Osgood, *The Day of the Cattleman*, (Chicago, Illinois: The University of Chicago Press, 1929).

4. The story of the accidental wintering of oxen on the plains of Wyoming was presented in the "Nimmo Report" and by John Clay in *My Life on the Range*, (Chicago, Illinois: Privately Printed, 1923). Osgood considered the often-quoted statement in the "Nimmo Report" that it was accidentally determined that cattle could winter in Wyoming as pure fraud, designed to dramatize the stock prospectus of ranching companies. However, Agnes Wright Spring *in* Maurice Frink, W.T. Jackson, and A.W. Spring, *When Grass was King*,

(Boulder, Colorado: University of Colorado Press, 1956), gives a detailed description of the event and places the date as 1857. A letter written by Alex Street, Superintendent of Wells Fargo & Company's freighting department at Cheyenne, Wyoming, and published by Dr. Hirman Latham, in *Trans-Missouri Stock Raising: the Pasture Lands of North America: Winter Grazing*, (Omaha, Nebraska Daily Herald Steam Printing House, 1871; reprinted by the Old West Publishing Company, Denver, Colorado, 1962). Latham reported he had wintered work oxen in Wyoming with no supplemental feed from 1856 to 1868. In about 1853, Seth E. Ward, Fort Laramie, began to winter cattle in the valleys of the Chugwater and Laramie rivers, according to Louis Petzer in *The Cattleman's Frontier*, (Glendale, California: The Arthur H. Clark Company, 1936).

5. J.A. Young, R.E. Eckert, Jr., and R.A. Evans, "Historical Perspective on the Sagebrush Grasslands," *Proc. Sagebrush Symposium*, (Logan, Utah: Utah State University, 1978). Information on livestock herding supplied by J.H. Robertson, Professor Emeritus, Range Science, University of Nevada, Reno, Nevada.

6. Dr. Latham published the first general appraisal of livestock raising on the open range. His pamphlet predated Joseph G. McCoy's *Historic Sketches of the Cattle Trade of the West and Southwest*, Ralph Bicher, ed., (Glendale, California: The Arthur H. Clark Co., 1940), by three years.

7. J.A. Larson, "The Winter of 1886–87 in Wyoming," *Annals of Wyoming*, (1942) 14:5–18. See also, Granville Stuart, *Forty Years on the Frontier*, D.C. Phillips, ed., Two Volumes, (Glendale, California: The Arthur H. Clark Company, 1925).

8. The story of drifting cattle was related by Osgood. The cattle drifting in the streets is from Robert Fletcher, "The Hard Winter in Montana 1886–1887," *Agric. History* (1930) 4(4):123–30. The newspaper accounts are from R.H. Mattison, "The Hard Winter and the Range Cattle Business," *The Montana Magazine of History*, (1951) I (4):5–22.

9. Bismarck weather data based on study of records by L.F. Crawford, *History of North Dakota*, (Chicago, Illinois: University of Chicago Press, 1931), Vol. 1, p. 484.

10. Harold McCraken, *The Charles M. Russell Book*, (Garden City, New York: Doubleday and Company, 1951).

11. Larson, "Winter of 1886–87 in Wyoming."

12. Frink, "When Grass was King." This citation emphasizes the first section in the volume.

13. This comparison was probably first published by Osgood. It has been noted by many authors, e.g., Frink, "When Grass was King." Larson in "Winter of 1886–87 in Wyoming," gives Chicago market prices for September 6, 1887, as $2.75–$3.45 per cwt and $2.40–$3.50 per cwt on November 8, 1887.

14. C.W. Gordon, "Report on Cattle, Sheep and Swine," *Supplementary to Enumeration of Livestock on Farms in 1880*, 10th Census of the United States, Vol. II, (Washington, D.C.: GPO, 1880). The story of wintering cattle in Ruby Valley is from Petzer, "The Cattleman's Frontier."

15. Ernest Antevs, *Rainfall and Tree Growth in the Great Basin*, (Washington, D.C.: Carnegie Institute, 1938). Most of the tree-ring data is obtained from the margins of the Great Basin.

16. C.B. Creel, *A History of Nevada Agriculture*, (Reno Nevada: Max C. Fleischmann College of Agric., Univ. of Nevada, 1964). See also Gordon, "Report on Cattle, Sheep and Swine." The Gordon report was based on a sample of seventy cattlemen who indicated losses of 20 percent.

17. Gordon, "Report on Cattle, Sheep and Swine." This report was prepared in the early 1880s, when the events of the winter of 1879–80 were relatively fresh in the minds of the stockmen interviewed.

18. C.W. Hodgson, "Idaho Range Cattle Industry," (Moscow, Idaho: Animal Science Department, Univ. of Idaho, 1948). A copy of this rare report was supplied by Idaho Historical Society.

19. Edna B. Patterson, Louise A. Ulph, and Victor Goodwin, *Nevada's Northeast Frontier*, (Sparks, Nevada: Western Printing and Publ., 1969). This regional history contains a great deal of source material collected by Mrs. Patterson during a lifetime in Elko County. It is further enriched by the other authors.

20. A.E. Granger, M.M. Bill, G.C. Simmons, and Florence Lee, *Geology and Mineral Resources of Elko County, Nevada*, Bulletin 54, (Reno, Nevada: Nevada Bureau of Mines, University of Nevada, 1957). Long-term precipitation averages are compiled in this bulletin. Newspaper accounts are from Patterson, et al., "Nevada's Northeast Frontier."

21. V.S. Truett, "White Winter of 89 and 90," *Nevada Magazine*, Vol. III (January 1948), pp. 5–9, 28. This article provides the basis for many of the comments about the midwinter period in Nevada. Quote from *Elko Weekly Independent*, 12 December 1889.

22. *Elko Independent*, 22 December 1889; 29 December 1889.

23. *Elko Independent*, 12 January 1890.

24. *Elko Independent*, 2 February 1890.

25. *Elko Independent*, 16 February 1890.

26. *Reno Evening Gazette*, 10 February 1890; 20 February 1890.

27. Byrd Trego wrote about the hard winter of 1889/90 in the *Idaho Republic* of 17 May 1928. This is one of the best first-person accounts of the hard winter of 1889–90.

28. Reports on the hard winter of 1889–90 were taken from J.D. Oliphant, *On the Cattle Ranges of the Oregon Country*, (Seattle, Washington: University of Washington Press, 1968). On page 284 of this volume, Oliphant provides newspaper citations for documentation of winter losses from each area mentioned.

29. E.F. Treadwell, *The Cattle King*, (New York: The MacMillan Comp., 1931).

30. H.O. Brayer, "The L-7 Ranch: An Incident of the Economic Development of the Western Cattle Industry," *Annals of Wyoming*, XV (January 1943):5–37.

31. Adelaide Hawes, *The Valley of Tall Grass*, (Bruneau, Idaho, Private Printing, 1950). The author confused the winter of 1889–90 with the winter of 1888, but the volume contains much valuable source material.

32. *Elko Independent*, 16 March 1890. Many knowledgeable residents of Elko County question the often-quoted story of stepping from carcass to carcass along the Marys River. The Marys River is not 100 miles long and many people believe the story refers to the Salmon River and Sparks-Tinnin range.

33. *Elko Independent*, 2 March 1890.

34. C.S. Walgamott, *Six Decades Back*, (Caldwell, Idaho: The Caxton Printers, 1936). This is believed to be the published origin of this statistic which Walgamott probably obtained from the folklore, newspaper accounts, and word-of-mouth accounts of the era.

35. *Harper's Weekly*, 20 June 1902, Vol. XLVII:1017–34. Hereafter cited as *Harper's Weekly*, "Nevada."

36. This version of the meeting was first published by Walgamott, "Six Decades Back." Tinnin did go to Nebraska where he continued to raise Longhorns; see N.S. Yost, *The Call of the Range: The Story of the Nebraska Stockgrower's Association*, (Denver, Colorado: Sage Books, 1966).

37. The $45,000 price is for one-half or one-third interest in a $1,000,000 stock company. The incorporation papers were filed with the Office of the Secretary of State of California, February 12, 1891.

38. L.R. Harris, "The Cultural Ecology of Sheep Nomadism: Northeastern Nevada, 1870–1972," Ph.D. diss., Yale University, New Haven, Connecticut, 1974.

39. Jerry Budy, and J.A. Young, "Historical Uses of Pinyon Juniper Woodlands," *Forest History* (1979) 23(3):113–21.

CHAPTER
8

WATER: THE FINITE RESOURCE

The winter of 1889–90 drove home a lesson that forage had to be conserved for wintering cattle on most of the sagebrush ranges of the Intermountain area. Where was the necessary hay to be produced? Most of the native plant communities were not suited for harvesting hay. The density of grasses and the total biomass of herbage produced were not suitable for hay production and they contained shrubs that were too woody for mowing.

Irrigation was only one way to produce forage in sufficient quantity to make harvesting hay economically feasible. Early ranchers such as Jasper Harrell had nature as a model to follow. Each spring when the snowmelt occurred, streams such as Goose Creek overflowed their banks and flooded the natural meadows including the Winecup field of Harrell's.[1]

Irrigation is such a commonplace thing in the western United States that it is difficult to comprehend that it was not always a part of American agriculture. The Mormon settlers in Salt Lake Valley are generally credited as being the first Anglo Americans to raise crops under irrigation in the western United States. On July 25, 1847, the day after the arrival of the first small group of Mormon pioneers in the Salt Lake Valley, a special service of thanksgiving was held. Orson Pratt was the principal speaker. He developed a prophetic vision of the parched desert land on which he stood. His text was from the words of Isaiah, "The wilderness and the solitary place shall be glad for them, and the desert shall rejoice and blossom as the rose."[2]

Obviously, the Mormons were not the first to raise irrigated crops in the western United States. Sophisticated Indian cultures of the American Southwest had developed irrigation systems that flourished and disappeared centuries before the arrival of the Mormons in Salt Lake Valley. Among the tribes of the Great Basin, Steward reported that Shoshone family units sometimes diverted ephemeral streams to increase grass seed production.[3]

Although the Mormon settlers are often credited with the conceptual model of modern irrigation in the Far West, a second center of development was the gold-mining industry of California. Virtually all forms of placer mining required the diversion of water. Hydraulic mining involved the diversion and long-distance transfer of water so sufficient head could be obtained to power the hydraulic giants. A technology developed for diversion dams, flumes, and inverse siphons could be transferred to irrigation works in other areas. In California many of the water diversion works that were developed for mining were converted into irrigation works with the depletion or regulation of hydraulic placer mining. This technology was spread to other territories by experienced miners. David Wall gained experience in the California goldfields and is often credited with developing irrigation systems in the Platte River Valley of eastern Colorado, independent of any Mormon influence.[4]

Not only was it necessary to develop the structures and agronomic principles necessary to raise crops under irrigation, it was also necessary to evolve laws governing water rights. The Mormon settlers had the structure of a church-oriented society to initially govern water policy. In the early Mormon settlements each head of a family was given 1.25 acres in town. Farmlands were given to settlers in ten- to eighty-acre tracts depending on distance from town. Those who followed occupations in town were given five-acre garden plots. In order to raise crops, the Mormons had to dig for irrigation water to each plot. The ditches were owned in common.[5] For the rest of the West, water law generally evolved from mining laws. In the eastern United States during the early nineteenth century, water, as the source of power for mills, was recognized as a legal use of the resource. After water falling over the mill wheel provided the energy to run the mill machinery, the water returned to the stream virtually in the same amount that was diverted from the millpond. In the West, water used in mining or in irrigation seldom returned to its natural channel in undiminished quantity.

Water law doctrines in all states of the United States are based on either the Riparian or the Appropriation principle. The Riparian principle is the right to use water, which is a real property right for most purposes, based on the ownership of land next to, or contiguous to, surface water. Under the common-law Riparian Doctrine, the right to use water is inseparably annexed to the soil. Use does not create the right, and misuse does not destroy or suspend the right. All states east of the 98th meridian except Mississippi and Florida follow the Riparian

Doctrine.[6] All states west of the 98th meridian, as well as Mississippi and Florida, recognize the Appropriation principle. Under this principle, water right is based upon beneficial use of the water. The first to appropriate and use water has a right superior to the rights of later appropriations.

Nevada and other western semiarid states were granted control of their natural waters for citizen appropriation, for irrigation and stock water, by the federal government. Legal difficulties over water started with the discovery of the Comstock Lode. The waters of the Carson River provided power for stamp mills for ore reduction in the Dayton area; transportation for wood products from the forests of the Sierra Nevada to the mines and mills; and irrigation water for the ranchers of the Carson Valley. The conflicts and resulting litigation that developed among these users is well-documented by the publications of Grace Dangberg and John Townley.

The 1889 Nevada legislature provided a means of determining individual water rights. The act, designed to regulate the use of water for irrigation and for other purposes, was modeled on Colorado law. The statute imposed a self-regulating system for determining water rights. The law divided the state into seven irrigation districts by major drainage basins, created water commissioners, and gave them authority to decide individual water entitlements within their districts. It decreed that all rights to water were to be filed with each county recorder by September 1, 1889, reserved unappropriated water to the state, and prevented enlargement or erection of irrigation works without express permission from the appropriate water commissioners. Following passage of the water-rights law, landowners began to file their claims to irrigation water with the various recorders. Individual claims were commonly exaggerated and far exceeded the ability of most streams' supplies.[7]

The first major attempt by the Nevada legislature to regulate water for irrigation was destined to have a short and rocky life. The years of 1888 and 1889 were extreme drought years in the Great Basin. The crucial legal conflict took place in Humboldt County, Nevada, where Judge A.F. Fitzgerald heard the suit entered by P.N. Marker, et al. against some 540 Humboldt River Valley irrigators. Drought conditions during 1889 forced Lovelock farmers to bring a common action under the Marker suit asking all Humboldt rights to be determined. Upstream ranchers in Elko County argued that settlement of individual water entitlements on the Humboldt could not be attempted by the court since the 1889 water control statute was unconstitutional. Judge Fitzgerald agreed, so the basic point at issue, Riparian versus prior Appropriation rights, was never considered. The 1893 Nevada legislature repealed the now-defunct 1889 water law and failed to enact anything in its place. The Nevada legislature next passed laws establishing the procedures for registering water rights in 1905. Water already appropriated for beneficial use at the time of the act was recognized as vested rights. The act provided that future appropriations were to be made by the state engineer. The water-rights law in Nevada did not specify how long people with vested rights had to record these rights.[8] Strange as it may seem, some ranchers, whose very livelihood depended on irrigation, delayed filing on vested rights that dated from the 1870s until the 1920s and early 1930s. The problem with the delayed filing on vested rights was that the dates, locations, and amounts of original appropriations were based on memory of individuals who appropriated or witnessed the appropriation of water as much as fifty or sixty years previous to the filing.

Nevada, along with some seven other mountain states, follows the Appropriation Doctrine exclusively. This doctrine is well-suited to areas that require the consumptive use of water. A consumptive use is one that takes on a substantially larger quantity of water out of the stream system than is returned after the water is used. The owner of an appropriation water right is entitled to use a stated amount of water even if such a use means that other would-be users with later priority dates are deprived of water. Most states date an appropriation water right from the first date water was put to a beneficial use. Usually credit is given when construction is begun on the diversion works.

According to federal court records, the Winecup fields of Jasper "Barley" Harrell were first artificially irrigated May 1, 1875, and at that time only 100 acres were irrigated. In 1886 irrigation was extended to an additional 125 acres and in 1900 an additional 100 acres were irrigated on a ranch with thousands of acres of sagebrush rangeland. The Rancho Grande, which was Sparks's favorite ranch in the Intermountain area, had a similar sequence of irrigation development with 300 acres developed in 1883 and 150 acres added by the turn of the century.

The Salmon River or Salmon Falls Creek originates and extends for many miles in Nevada, running in a generally northerly direction until it crosses the northern boundary line into the state of Idaho and discharges into the Snake River. The drainage system belongs to the Snake River hydrographic basin rather than the Great Basin. The stream had to fight to escape the confines of the Great Basin. North of the Vineyard and Hubbard ranches the intrusive granite rocks that support the mineralized area around Contact were almost too indurate for the channel to wear through; again, north of San Jacinto, basalt flows reach down from the highlands and threaten the course of the river. The major portion of the waters of the Salmon River are supplied by the O'Neil Basin and Shoshone Creek. The headwaters of O'Neil Basin are the 10,000-foot-plus peaks of the Jarbidge area. The river leaves O'Neil Basin through a narrow gorge and confluences with Jakes Creek, which flows in from the south at the Vineyard Ranch. The river flows northerly from the Vineyard to the state line, and is joined by Trout Creek and Shoshone Creek a few miles south of the state line. Shoshone Creek, a principle tributary, rises in Idaho, and flows southerly, crossing the state line east of present Jackpot, Nevada. It then flows in a generally westerly direction and joins the main stream at the lower end of the San Jacinto Meadows. The Salmon River was, by far, the largest stream in Sparks's ranching empire and the major base of irrigated hay production. Eventually over 10,000 acres of irrigated land were developed along the Salmon by Jasper Harrell, Sparks and Tinnin, and Sparks and Harrell.[9]

According to the District Court records, irrigation works were started by Jasper Harrell on the Salmon River in 1873. The first irrigation season was 1874. The East Bird's Nest Slough Ditch, the West Boar's Nest Ditch, the Salmon River Sloughs, and the Harrell or Big Ditch brought over 2,000 acres under irrigation. In 1878 another 800 acres were brought under irrigation, mainly on tributary streams such as Jakes Creek to the south of the Vineyard Ranch. For the next ten years, the irrigation system remained relatively stable while the number of livestock run on the ranges expanded by at least a hundredfold. Sparks-Tinnin were not very active in enlarging the Salmon River irrigation system. In 1887 the Willow Spring or Dry Creek Ditch brought about 700 more acres under irrigation, but this was the only project developed while John Tinnin was Sparks's partner.

After the winter of 1889–90, the need for additional hay production was easily recognized. When Jasper and Andrew Harrell returned to the Sparks Company, they started new, higher elevation ditches and extended the existing ditches. By 1894 about 10,000 acres were brought under irrigation. A major part of this addition was the Highline Ditch which added 2,500 acres.[10] By 1900 Sparks maintained that his network of ranches produced 15,000 tons of hay per year, but more importantly, his Great Basin wildrye meadows produced the equivalent of 100,000 tons of standing cured hay. This would indicate Sparks-Harrell had fenced 50,000 acres of naturally occurring Great Basin wildrye communities. John Sparks enjoyed bragging on his meadows or Salmon Falls Creek which was fourteen miles long and ran near San Jacinto. John Sparks's idea of a suitable photograph of his ranch was to be in his fancy buggy with a flashy team of horses, and Great Basin wildrye higher than the buggy wheels.

In the diversion of irrigation waters, ranchers followed the procedures developed by placer miners. A rock dam was built across the stream at the point of diversion. It usually was not a structure of cut stones, only rock piled and lodged with whatever timber, cottonwood, or aspen logs that were available. Often the first spring flood would wash out the dam and rock would have to be hauled in wagons to repair the dam before the irrigation season could start.

The first ditches had their origin on the land claims to be irrigated. Water was diverted on the upstream end of the claim and spread as far as the grade would allow. Most early land claims were on streams located in valleys. The simple diversion of water on their own land allowed the lowlands along the streams to receive supplemental irrigation water. To get water up to the first stream terrace above the bottom lands, it was necessary to put the point of diversion further upstream so the necessary elevation could be maintained with the ditch.

In the Bruneau River Valley of southern Idaho, John Baker dug the first ditch in 1876. Each rancher, large and small, with land along the Bruneau River, established ditches through the late 1870s and early 1880s. It soon became apparent that to bring the upper terraces under irrigation, much longer and larger ditches would be required. The soils of the first terrace above the stream bottoms usually were deep, finely textured, and the great size of the sagebrush indicated it could be productive. It was soon found that irrigation works were necessary on a larger scale than the individual settlers could afford to build. In the

1880s many irrigation companies were formed throughout the West. These companies varied in their design and complexity. Some were informal arrangements among groups of ranchers who shared work on a ditch. Some companies sold shares of stock, thus obtaining money with which to build the dams and ditches necessary to make the system work.[11] The acquisition of water rights and the development of irrigation systems were among the stated purposes of the incorporation papers of the Sparks-Harrell Company. The company undertook its own water development.

Teams drawing Fresno scrapers were used to help construct ditches through alluvium. Rocky slopes and cuts were pick, shovel, and blasting operations. Construction of ditches and irrigation structures was largely done by hand labor. The cost of digging the Orr Ditch from the Truckee River varied from $0.75 to $10.00 per rod (16 feet) depending on the nature of the ground through which the ditch was being constructed.[12]

Irrigation water was flooded out onto the new land often without bothering to remove the sagebrush. The native desert shrubs could not stand flooding and soon died with irrigation. Cattle wintered on the newly irrigated lands trampled the brush into contact with the soil and it soon disappeared. The first terrace often contained small amounts of wet meadow species scattered under the shrubs. The native sedges and rushes rapidly increased on the newly irrigated soils and soon the first terrace sites supported plant communities remarkably similar to the flood plain.[13]

Many ranchers made attempts to seed the newly irrigated lands. A familiar sight was a mounted rancher crossing his new field with his winter felt hat pulled down hard against the March winds as he dipped into a gunnysack of grass seed, hooked over the saddle horn, and broadcast the seeds to the winds. Favored grass species included orchard grass, timothy, and redtop. Many of the early ranchers were experienced with agriculture in humid environments and these species fit their conception of what a wet meadow hayfield should contain. All of these species naturalized in the Intermountain area, but they rarely dominated what became known as the native hayfields.

The land usually was not leveled before irrigation. Natural drainage ways and swales gradually sodded in and lost some of their relief. Irrigation of these undulating meadows became an art. In the spring, wagonloads of manure were hauled to the fields to make strategically located dams. Manure from work and saddle horses that were kept in during the winter provided a convenient construction material that was in oversupply in the corrals and horse sheds. Tremendous heads of water were used in the spring. The ranchers called it "taking the frost out of the ground." The low places might be four feet deep in irrigation water and the high places took care of themselves. Old-time ranches that obtained their irrigation waters from cooperative farmers' ditches are often advertized for sale as having "free" irrigation water. This is opposed to formal irrigation districts where a fee is charged for the water. The free water is very much a misnomer because the maintenance of ditches and diversion structures has to be done by the ranch owner.

Mucking ditches in the spring became a familiar part of western ranching. Standing in gummed rubber boots in six inches of mud in the bottom of an irrigation ditch while cutting and lifting heavy sod out of the ditch was a long way from the "knight of the plains" concept of a cowboy. Mucking ditches was a spring social event when ranchers worked together on cooperative ditches. There were very few nineteenth-century ranch jobs where the crews had the time, were close enough together, and it was quiet enough for conversation. Mucking ditches was one of the rare jobs where the crews could visit. This usually ended in the stock argument of the bunkhouse crew which pitted the tough, withdrawn, old-timers against the brashness of younger workers. Bragging on prowess in the houses on the wrong side of the tracks in Elko was greeted by a chorus of scoffing which in turn brought retorts of "it was too bad the old-timers could not even remember what they were missing."

A sharp shovel is required for cutting sod and the conversation of men toiling in the muck was punctuated by the harsh ring of a fine flat file milling the cold steel of a good shovel blade to a razor edge. More than one hernia was attributed to a cowboy-cum-ditch-mucker who overestimated the weight of a chunk of sod he could pitch out of a ditch.

The newly created native hay meadows were generally irrigated or, more correctly, flooded only once each season. When the low spots more or less dried out it was time to mow the forage for hay. On exceptionally wet years, the bottom lands sometimes failed to dry sufficiently to allow the hay to be cut and cured. This delay caused by high water was especially common along the Humboldt River.

The practice of the early ranchers, who had the oldest priority water right, of raising native meadow hay under flood irrigation, was bitterly attacked by agriculturists and sociologists. The waters of a given stream were appropriated by successive diversions until all possible water was placed in use. The climate of the Great Basin is highly variable. There is really no such thing as an average year in terms of an actual pattern or amount of precipitation. Averages are products of statistical treatment of extremes, and droughts are an all too often part of the Great Basin environment. When droughts occurred, the ranchers with the oldest water rights diverted water as usual and flooded their meadows. Ranches with later rights often had to do without irrigation water.

In a written deposition for the Nimmo Report, Merrill described the water conflicts that were prevalent in the early 1880s: "A certain number of farmers unite and build a ditch from a river to irrigate their lands; an adjoining set build another for their lands, and so on until several ditches are in operation. The ditches being under separate management, conflicts ensue and damage suits follow, as one company fails to take care of its waste water, allowing it to run on the farm of a stockholder of another company, thus wasting water that might be used to an advantage in another direction."[14]

Water rights, like all property rights, are subject to qualification and to regulation by the state. There is no such thing as absolute ownership. Society has the common-law right and relies on the law to settle conflicts over water. However, the waste concept has rarely been introduced to water law to force efficient irrigation practices. Courts have been faced with this question of how to increase efficiency without infringing on the basic property right connected to water appropriations for irrigation. In much of the Intermountain area, failure to increase efficiency in irrigation had social ramifications throughout the area. The wild flooding practices of irrigating native meadows for hay production were perpetuated and technological agronomic farming was restricted because the old-time ranches had the priority water rights. This kept northern Nevada in a cow-country economy.

In order to irrigate lands farther from and higher in elevation than the flood plain bottoms, it was necessary to dig miles of ditches and have the diversion point located long distances from the area to be irrigated. The Nimmo Report indicated that in 1884 there were 800 irrigation ditches in Nevada, aggregating about 2,000 miles in length and irrigating 150,000 acres. Obviously, a lot of ranchers were dependent on lifeline ditches strung out for several miles.

Soils of arid regions are different from those of humid areas. Having developed under low rainfall and scant vegetation, the soils have been weathered and leached less and contain smaller quantities of organic matter than their humid-region analogies. Desert soils usually contain appreciable quantities of lime, magnesium, and other mineral elements in unweathered and precipitated minerals. Soils of humid regions are generally leached of these ingredients. Arid-region soils are normally alkaline in reaction and in some situations are being enriched in basic minerals through irrigation. Humid-region soils tend toward acidity.[15] These distinctive characteristics of arid soils must be considered in managing them for crop and forage production. Farmers on these soils escape the problems of acidity and liming, but have the different problems of excess lime, alkalinity, and soluble salts.

Mormon pioneers arrived in Salt Lake Valley on July 24, 1847. In order to plow the dry and baked ground, they diverted the waters of City Creek for irrigation. When this water was put on land which had developed under fifteen inches of rainfall, a new cycle of nature began. The fifteen-inch annual rainfall had fallen on soil where the potential evaporation was sixty inches. Consequently, leaching had occurred only during exceptionally stormy periods and these infrequently percolating waters had not reached any but the highly soluble products of rock weathering from the soil. Suddenly with irrigation the soils' environment was changed from an arid to a humid state. The natural balance between precipitation and evaporation was reversed.

The first unfavorable result noted from irrigation was the gradual building up of groundwater in some areas. The first lands to be irrigated were the low-lying lands of fairly level topography. Later water was diverted on to higher benchlands. The drainage water seeped into the lands below, reducing aeration, and bringing in excess salts until, in many cases, these once highly productive lands were reduced to low-production levels.

No doubt, the most severe problem arose from too much water rather than too little. But the proper balance among water, soil, and crops is difficult to determine and maintain. Excessive use of water leaches many plant nutrients from the soil and leads to problems of drainage, aeration, and salt. To economically use water leads to deposits of undesirable minerals in the soils irrigated.

Irrigation waters differ from natural precipitation in that they usually contain larger amounts of dissolved impurities. Sodium is the most common injurious ingredient of irrigation waters. Sodium is not more toxic to plants than calcium or any other element, but if the concentration of sodium exceeds calcium and magnesium, the sodium will be absorbed to the clay particles displacing the normal high proportion of calcium. This causes the clays to disperse and loose their normal structure. Farmers call such soils "slick spots." Such soils have virtually no drainage and are very poor soils for plant growth. As early as 1880 farmers in Nevada were experimenting with the addition of gypsum to counteract saline/alkaline soils.[16]

The ranchers and their irrigators were not trained engineers. They took all the water they could get and that the ditches would hold. This led to many conflicts when the waters of given streams were completely appropriated. Thousand Springs Creek Valley was a valuable part of the Sparks-Harrell ranching empire. There are approximately 15,000 acres in the valley that can be readily irrigated with water diverted from the streams or the many springs. The available water averages 18,000 acre-feet annually. An acre-foot of water is the amount necessary to cover one acre of land to a depth of one foot. Most of the irrigable land occurs on the branches of the stream near Montello, Nevada, where Sparks's Gamble Ranch was located.[17]

Snowmelt in the high mountains of the Thousand Springs Creek watershed produced maximum seasonal streamflow during April, May, and June. Low water occurs from August through October. The valley floor of Thousand Springs Creek receives five to eight inches of annual precipitation. The mountains that form the headwaters of the stream receive fifteen inches of precipitation. Irrigation is totally dependent on the runoff from the high-elevation precipitation. The condition and quality of the watershed eventually determine the tenure, quality, and quantity of irrigation developments. The quantity and density of plant cover on the soils of the watershed contribute to the rate of runoff and the amount of siltation as soils are eroded. Overgrazing and destruction of the vegetation cover eventually lead to floods. During the period from 1900 to 1920 tremendous floods came down Thousand Springs

Creek to pass through Montello and out onto the salt flats of pluvial Lake Bonneville. The floods occurred well after the time scale of this narrative, but the seeds of these disastrous floods were planted in the grazing practices of the 1870s.

The necessity of irrigation to produce hay crops made it obvious that some special land policy had to be developed to encourage and accommodate irrigation of western lands. The Desert Land Entry Act of March 3, 1877, allowed entry on 640 acres of land. This was four times the amount of land allowed under the preemption or homestead methods of land entry. Eastern congressmen fought long and hard to block passage of this act, while representatives of western states stressed that the cost of developing irrigation systems made the larger acreage necessary. After many complaints about illegal land entry under the Desert Land Entry Act procedure, the law was amended in 1890 so that a maximum of 320 acres could be obtained.[18]

The original law of 1877 was restricted to the states of California, Nevada, Oregon, and the territories of Arizona, Dakota, Idaho, Montana, New Mexico, Washington, Utah, and Wyoming. It was amended in 1891 to include Colorado. The stated purpose of the Desert Land Entry Act was "to encourage and promote the reclamation, by irrigation, of the arid, and semiarid lands of the western states." It was assumed that settlement and occupation would naturally follow when the lands had been rendered more productive and habitable.

To be eligible for Desert Land Entry, the land must be irrigable and not capable of producing a crop without supplemental irrigation. An exception to the crop rule was alternate year grain/fallow systems of farming. Application for Desert Land Entry had to be accompanied by evidence of water rights already acquired by appropriation, contract, or purchase of a right to permanent use of sufficient water to irrigate and reclaim all of the irrigable portions of the entry. Desert Land Entries had to be as nearly square as possible even when entered on unsurveyed land. This specification was a major problem when irrigable acres were restricted to long, narrow strips along streams. The entire entry had to be in one piece and not in disjunct tracts.

There was probably more fraud connected with Desert Land Entries than any other method of disposal of the public domain during the late nineteenth century. In this connection it is worthwhile to revive the procedures used for establishing the final proof for Desert Land Entries.

The claimant had to name four witnesses for the final examination. The examiner chose two of the four witnesses. These witnesses were persons who were familiar, from personal observations, with the land in question and what had been done toward reclaiming the land.

Final proof had to show specifically the source and volume of the water supply, the number, length, and capacity of ditches on each of the legal subdivisions. Witnesses had to state if they saw the land effectually irrigated and the different dates on which they saw it irrigated. As a general rule, actual tillage of one-eighth of the land had to be shown. The original law required that all the 640 acres had to be irrigated, but this was practically impossible. It was not sufficient to show a marked increase in forage production from native grasses with irrigation. An actual crop had to be planted and irrigated.

Despite the fact that the idea for Desert Land Entries originated with experiments in Lassen County, California, and was specifically designed for Nevada, this method of disposal of public lands was never popular in the Great Basin. In Nevada 611,320 acres were entered under this act and 143,000 were finalized. There was plenty of irrigable land in the Great Basin that would respond to the application of water. Unfortunately, there was not sufficient water to irrigate more than a small fraction of this available arable land.[19]

There are three major significant factors associated with nineteenth-century development of irrigation in the sagebrush/grasslands environment. First, the available forage base was greatly increased and this forage could be conserved for winter hay feeding. Secondly, the successful production of irrigated crops in the Intermountain environment provided the germ of the idea that developed into the government-sponsored land reclamation projects of the early twentieth century. Finally, the incorporation of irrigated farming with range livestock operations contributed to the further grounding of the cowboys. The wild horseman of the plains lost considerable glamour but became more functional by standing in a muddy ditch leaning on a shovel.

Notes

1. The development of water rights on Goose Creek is reviewed in the records of the District Court of the United States, for Idaho, Southern Division, for the case of Twin Falls-Oakley Land and Water Company and Oakley Land Company vs. Vineyard Land and Stock Company, August 14, 1915. Transcript on file with Nevada State Engineer's Office, Carson City, Nevada. Jedediah Smith visited Salmon Falls Creek in 1826, but failed to find beaver. Milton Sublette and a party of trappers from the Rocky Mountain Fur Company trapped beaver along Goose Creek in 1837. See D.L. Morgan, *Jedediah Smith and the Opening of the West*, (Lincoln, Nebraska: University of Nebraska Press, 1953).

2. D.W. Thorne, *The Desert Shall Blossom as the Rose*, 10th Annual Faculty Research Lecture, (Logan, Utah: The Faculty Association, Utah State Agricultural College, 1951).

3. One of the early books on irrigation with reference to irrigation by Mormons is by F.H. Newell, and D.W. Murphy, *Principles of Irrigation Engineering*, (New York: McGraw-Hill Book Co., 1913). Comments on Indian irrigation are from J.H. Steward, *Basin-Plateau Aboriginal Socio-political Groups*, (Washington, D.C.: GPO, 1938).

4. Art Steinel, *History of Agriculture in Colorado*, (Fort Collins, Colorado: State Agricultural College, 1926).

5. A.H. Sanford, *The Story of Agriculture in the United States*, (Boston, Mass.: D.C. Heath and Company, 1961).

6. T.A. Garrity, and Elmert Nitzschke, Jr., *Water Law Atlas*, (Socorro, New Mexico: New Mexico Chapter of the Federal Bar Association in cooperation with the State Bureau of Mines and Mineral Resources, New Mexico Institute of Mining and Technology, Cir. 95, 1957). This atlas is an excellent and simple presentation of the basic principles of water law.

7. Grace Dangberg, *Conflict on the Carson*, (Minden, Nevada: Carson Valley Historical Society, 1976). See also, E.O. Wooton, *The Public Domain of Nevada and Factors Affecting its Use*, (Washington, D.C.: GPO, Bull. No. 301, U.S. Dept. of Agric., 1932).

8. J.M. Townley, "Reclamation in Nevada, 1850–1904," Ph.D. diss., University of Nevada, Reno, Nevada, 1976. See also, Myron Angel, ed., reproduction of 1881 edition of Thompson and West's *History of Nevada* with illustrations and biographical sketches of its prominent men and pioneers, (Berkeley, California: Howell-North Publishing, 1958).

9. "Biennial Report of the State Engineers," State of Nevada, June 30, 1939 to July 1, 1940, (Carson City, Nevada: State Printing Office, 1940).

10. Records of the District Court of the Fourth Judicial District of the State of Nevada in and for the County of Elko. Judgement and decree of the matter of the determination of the relative rights in and to the water of the Salmon River, March, 1922. See also R.I. Fulton, "Camplife on Great Cattle Ranges in Northern Nevada," *Sunset* Vol. V (No. 3, July 1900):111–18. Fulton is often misquoted to indicate that Sparks harvested 100,000 tons of hay.

11. Adelaide Hawes lists the priority of all the water rights on the Bruneau River in her book, *The Valley of Tall Grass*, (Bruneau, Idaho: Private Printing, 1950). The history of irrigation districts from Sanford, "History of Agriculture."

12. The Alvaro Evans family papers are on file at the Nevada Historical Society, Reno, Nevada. The item is dated June 1879.

13. C.A. Brennen, assisted by C.E. Fleming, G.H. Smith, Jr., and M.R. Bruce, *Cost of Producing Hay in Nevada*, (Reno, Nevada: Agric. Expt. Sta., University of Nevada, 1932).

14. Nimmo Report, 1884–1885. Report from the Chief of the Bureau of Statistics in response to a resolution of the House calling for information concerning the range and ranch cattle traffic in the western United States and territories, (Washington, D.C.: GPO, Ex. Doc. No. 267, 48th Congress, 2nd Session, House of Representatives, 1889).

15. Comments on desert soils and water quality are largely based on Thorne, "The Desert Shall Blossom as the Rose."

16. *Reno Evening Gazette*, 16 October 1880.

17. R.E. Rush, *Water Resources*, Reconnaissance Series Report 47, Dept. of Conservation and Natural Resources, (Carson City, Nevada: State of Nevada, 1968). Water resource appraisal of Thousand Springs Valley, Elko County, Nevada.

18. B.H. Hibbard, *A History of the Public Land Policies*, (New York: The Macmillan Co., 1924); and, *Statutes and Regulations Governing Entries and Proof under the Desert Land Laws*, (Washington, D.C.: GPO, General Land Office, Cir. No. 474, Dept. of Interior, 1916).

19. Hibbard, "A History of Public Land Policies."

III
THE
LAND IN
TRANSITION

THE LAND IN TRANSITION

The cold deserts are lands of extremes. Bitter cold and snow are followed by burning heat and drought. The environment of the Great Basin had to be modified to permit the raising of cattle on a sustained basis. Such modifications included the painfully slow digging of irrigation ditches to bring the land into production for crops of hay. To come to the sagebrush/grasslands and raise cattle, ranchers had to be brave, and to survive in such enterprises, they had to learn that water runs downhill—most of the time.

CHAPTER
9

MAKING HAY IN THE GREAT BASIN

The lament of a cowboy in the 1890s was, "Cowboys don't have as soft a time as they did. I remember when we sat around the fire the winter through and didn't do a lick of work for five or six months of the year except to chop wood to keep us warm."[1] With the advent of hay production it was irrigation and haying in the summer and feeding the darn stuff in the winter. All of the old riding, roping, and branding was sandwiched in the spring and fall and as time permitted during the rest of the year.

Haymaking became the focal point of livestock production in the Great Basin. This was caused by the bitter experiences with severe winters, the gradual over-utilization of the forage resource, and finally, the development of supplemental irrigation for meadowlands. Ranchers spent half of each year culturing, harvesting, and storing hay, and the other half feeding hay to their wintering brood cow herds. The generally accepted rule of thumb in the Intermountain area has been that at least one ton of hay is necessary to winter each brood cow.

The economic, social, and ecological changes that were the direct and subtle effects of making hay changed the life-style of the residents and grossly influenced the cold deserts of western North America. Hay production converted the range livestock industry of the sagebrush/-grasslands from an extensive enterprise with minimum labor to a labor-intensive enterprise. The feeding of hay increased labor requirements during the winter, but the maximum labor requirements were highly seasonal with the peak during the summer haying.

The hay hands were largely a bachelor society, so the migratory aspects of their employment did not include the horrors of child labor, lack of educational opportunity, and self-perpetuation associated with twentieth-century migratory farm workers. The ranch bunk houses with their full allotments of bachelors contributed to making Nevada the most "male" state in the union, with more than twice as many men as women, and the smallest proportionate number of women and children.[2]

159

The ranchhand created unique social and political conditions in the sparsely populated areas of the sagebrush/grasslands. A characteristic sight in Nevada from the 1890s to the 1940s was in the words of social reformer Anne Martin, "the groups of roughly dressed men aimlessly wandering about the streets or standing on the street corners of Reno, Lovelock, Winnemucca, Battle Mountain, Elko, and Wells." They were in from the ranches with money to spend. Liquor, gambling, and women were the outlets for these sprees. Each small rural town shamelessly flaunted a red-light district, usually surrounded by a high board fence, erected as an illusionary barrier to protect the children of the townspeople, and popularly known as the "stockade."

The migratory nature of the ranchhands and their almost complete lack of civic awareness removed a large portion of the potential electorate from the political process. This contributed to the relative backward attitude of the Nevada legislatures toward social reform. Nevada was one of the last western states to adopt the women's suffrage amendment in the twentieth century. The very small voting electorate contributed to the control of Nevada by various special-interest groups.[3]

Social reformers, such as Anne Martin, bitterly attacked the injustices, real and imagined, of the seasonal employment of ranchhands. Rarely did ranchhands think of themselves as a discriminated-against minority. Ruggedly independent, they viewed themselves as masters of their own destiny, just as the ranch owners, whose market was controlled by relatively few meatpackers, treasured their self-image of independence.

A safety valve in rancher-ranchhand labor relations was the practice of "going down the road." The labor supply for ranching was usually insufficient, at least seasonally. If a ranchhand became dissatisfied on one ranch he could always go down the road and find a job. Some employees spent a lifetime at a single ranch, while others regularly went down the road. If an incident did not justify leaving, the chronic quitter would invent reasons for going down the road. Within a single valley or ranching area, ranchhands often made a circuit from ranch to ranch, eventually coming back to their original employers and forgetting the incident which sent them down the road. The circuit required a few months to twenty years depending on the intensity of the incident that sent the ranchhand down the road and the memory of the employee and the employer. Through the ups and downs of ranchwork, the men gloried in the fact that they did not have a union to tell them what to do.

More than one ranchhand went on a roaring drunk on the wrong side of the tracks in Elko, woke up as the member of a haying crew forty miles from the nearest county road, and claimed he was shanghaied. There are ranches in Nevada today that hire hands with the understanding they will be picked up and delivered to town after a month's work and no personal transportation is allowed at the ranch. If the ranchhand is going to go down the road before the month is up, he will do so on foot and it is fifty miles to town.

The ranchhand, who was part cowboy and seasonally a hay hand, ditch mucker, fence builder, was employed as a fixture of agriculture from the 1890s to 1942. This was not a short-term phenomenon. The ranchhand society lasted for fifty years in the sagebrush/grasslands where the grazing of domestic livestock is little more than a century old.

Any broad-brush portrait of the ranchhand is doomed to failure. Excessive use of alcohol, limited education, skill in working with horses and his hands, but without the marketable skill of a carpenter or blacksmith, all fit as generalities. A short spree of outrageous behavior that flaunted accepted morality provided both an identity and an outlet from weeks of poor living conditions and hard work. H.L. Davis, probably drawing on his own life as a drifter, described the feelings of a bachelor cowboy riding into an eastern Oregon town in the early evening. Davis expressed the contempt of the cowboy for the families safe and snug behind lamp-lit windows and took pride in the stares of disapproval that followed him down the street.[4]

Not all hay hands were bunkhouse bachelors. The seasonal nature of the work blended with the activities of small-time miners and prospectors, woodcutters, homesteaders, and many Indians who worked in the hay for a cash income to supplement their other activities. Haying was the first work experience for many school boys both from rural areas and from the towns of the sagebrush/grasslands. Almost every description of haying operations includes descriptions of activities performed by boys.

Some of the hay crews were railroad tramps. Some seasonal hay hands wintered in the Mediterranean climate of California and rode the

Central Pacific across the mountains to work in the hay. Some would return to the same ranch every year. There were railroad tramps who only worked under the most desperate situations, if ever. Henry Miller, with ranches from California through Nevada to eastern Oregon, developed a special policy toward such tramps. His rules were as follows:

1. Never refuse a tramp a meal, but never give him more than one.

2. Never refuse a tramp a night's lodging. Warn him not to use any matches and let him sleep in the barn, but only for one night.

3. Never make a tramp work for his meal. He is too weak before and too lazy afterwards.

4. Never let tramps eat with the men. Make them wait and eat off dirty plates.

The Miller and Lux ranches became known as the "dirty plate route" by tramps. Henry claimed his policy toward tramps saved him millions in fire insurance premiums.[5]

There was a definite class structure among hay hands. As a small boy, J.A. Young learned this from water dippers. His grandmother's house had a well located on a side porch. The ranchhands used the well to fill their water bottles, usually gallon wine bottles covered with burlap sacks. On a porch post by the well there was a series of metal dippers. A ranch foreman explained to Young that the top dipper was for his grandmother and her house people, the next one down the pole was for the cook, foreman, and blacksmith. The next nail held a dipper for the seasonal hay hands. Lowest on the pole was the "happy" dipper. It was for the personal use of a mentally unbalanced hay hand named Happy for his perpetual smile. No one wanted to catch a case of the "happies" by drinking from his dipper.

The early Mormon settlers had a unique way of establishing rights on the common forage grounds. On the night that it was determined the hay was ready to cut, a community party was held. At midnight the men adjourned to the haying grounds and began to cut a swath around the hay they proposed to cut. They could have all the hay they could surround by daybreak. If they set their sights too high and failed to close their piece by sunrise anyone had the right to cut in their area.[6]

The making of hay did not suddenly start after the hard winter of 1889–90. Hay was always a part of Nevada's agriculture. Almost every ranch put up some hay for stock horses from the earliest days of ranching in the sagebrush/grasslands, but the quantity of hay produced to number of cows wintered was such that only a small portion of the stock could be fed for a short period of time. Some of the large ranches cut considerable hay before 1889–90. According to the Elko paper, Pedro Altube of Independence Valley planned to cut 8,000 tons of hay during the 1887 haying season.[7]

A few early hay ranches took favorable sites along the California Trail, but local markets were necessary prior to significant agricultural development in northern Nevada. These markets first appeared along the Humboldt mining district in 1860 and 1861. California-bound emigrants passing through the Great Basin recognized the natural haylands adjacent to such streams as Goose Creek, Thousand Springs Creek, and the Truckee and Carson rivers.[8]

With the huge number of draft animals that were required to move supplies to the mines and ore to the mills, a hay production industry developed to feed the animals. The environmental constraints of vast stretches of desert without sufficient forage to support draft animals by overnight grazing while on travel or freight trips across the state, and cold, snowy winters when little forage was available even in the irrigated valleys made the conservation of forage as hay necessary. Hay found a ready market with emigrants or teamsters. Demand for hay grew during the 1850s as western Nevada attracted settlers and then skyrocketed in the 1860s when mining supported thousands of draft animals. In western Nevada hay ranchers faced a crisis shortly after 1868 when the Central Pacific Railroad completed its line into Reno and reduced the many draft animals needed to haul supplies over the Sierra Nevada to the Comstock Lode. Hay is a bulky product that is difficult to compress. Stationary hay presses existed in the nineteenth century, but the degree of compression obtained was not great and the process expensive. The ranchers of the Truckee Meadows area specialized in producing hay for use at remote mining sights. Alvaro Evans shipped baled hay by railcar on the Central Pacific and then by narrow-gauge railroads to Austin or Eureka, Nevada. The bales from a stationary hay press were large, with fourteen bales constituting a carload.[9]

The hay farmer generally had to have a market relatively close to his hayfields. Obviously, not every hayfield could be close to mines, transportation centers, or logging operations where large numbers of draft animals required hay for forage. The answer was to take the animals to the hay; not draft animals but beef animals who could walk to the hay and then walk to the railroad for shipment to market.

Agriculture, during the three decades following initial settlement in what was to become Nevada, was perennially subordinate to mining. The preeminence of mining after discovery of the Comstock Lode, and subsequent extension of the mining frontier eastward throughout the state overshadowed the related growth of irrigated farming and livestock production. Production from the Comstock silver mines toppled after 1876, but desperate Nevada residents found that a thriving range cattle industry had developed in excess of what was necessary to feed the mining population. Based on the reports published by the state surveyor general, hay production in Nevada climbed from 65,900 tons in 1873, to 618,000 tons in 1900. In 1873, Elko County had 15,000 acres under cultivation for hay production with an average production of one ton per acre. By 1880, 16,000 acres were devoted to hay production in the county and production climbed to better than two tons per acre. The climb in production probably reflects the development of irrigation systems. After the hard winter of 1889–90 the hay acreage shot up to 239,000 acres in Elko County and production per acre dropped to about one ton per acre, a level of production that was maintained for the next half-century.[10]

Truckee Meadows became a center of hay production for the Comstock Lode area and, after the decline of mining, the meadows became a cattle-feeding area. The Central Pacific Railroad stimulated winter feeding in the Truckee Meadows by allowing cattle firms to ship animals to Reno and, after being fed hay during the winter, reshipped to California at the rate charged for a simple Nevada-to-California transfer.[11] During 1870 Reno hay ranchers advertised their crops in state newspapers offering to feed cattle at a per-animal rate or sell hay by the ton. The herds were shipped by rail or driven to the host rancher's land. Hay production reached 5,000 tons with only 1,000 tons finding market to draft animals. Without winter feeding, prices were expected to fall from the average of $25 per ton but the development of feeding stabilized prices at $20 per ton.[12]

Ranchers who sold hay to teamsters often sold their entire crop to one purchaser and then were paid as the feed was delivered during the following fall and winter. Many hay ranchers contracted as teamsters in the off-season. They had to have horses and wagons to put up hay so contract hauling was a natural extension. Contracts for hay and freighting were made on the basis of regular delivery to freighting companies, stage lines, and other ranchers.[13]

Early hay makers were not particularly choosy about what they cut for hay. David Griffiths listed among the important hay species alkali bullrush, cattail tule, and spike rush. All of these species inhabit saline or fresh water marshes and seasonal lakes.[14] To this list of unusual hay species we could add an occasional willow shoot and a rabbitbrush stem. Professor F. Lamson-Schriber, commenting on western hay species in 1883, suggested a brush scythe was a suitable implement for cutting many of the hay crops.[15] When questioned about the quality of his native hay one old rancher replied, "when the snow is on the ground all they want is something to chew—the deeper the snow, the harder they chew."

One of the most important native hay species was creeping wildrye. This species is the diminutive relative of Great Basin wildrye. It is a much smaller grass with abundant scaly rhizomes or underground stems. Early ranchers in Nevada knew the species as blue joint. Griffiths considered this species to be the most important grass in the Intermountain region. He compared its manner of growth to western wheatgrass on the high plains. Both western wheatgrass and creeping wildrye are rhizomatous species. Griffiths found magnificent stands of creeping wildrye along the Humboldt and Quinn rivers. It was growing on rich, non-alkaline, heavy-textured soils and when properly irrigated yielded two to two-and-one-half tons of hay per acre.[16] Creeping wildrye was the only native grass that Griffiths considered adapted to the wild flooding method of irrigation which left the low spots in the field covered with relatively deep water for extended periods. Of the introduced grasses, only redtop was sufficiently naturalized to warrant comment by Griffiths. He despaired at its failure to gain a dominant role in the hay meadows.

Willows are very much a part of hay meadows. Where stream gradients flattened in natural meadows, the streambeds tended to meander in S-shaped loops. In time, the streams cut through the loops leaving oxbows. These oxbows provided habitat for strips of willows. In late spring when the meadows were being flooded before hay harvest the herbaceous vegetation was so green it assaulted the senses made hungry for contrast by endless sagebrush gray. The willows, arranged in sensuous curves, proved a gray-green contrast to the pale green of wire-grass.

The introduction of a single forage species, alfalfa, changed livestock production and the economy of the entire Intermountain area. Because it was a legume that supplied its own nitrogen through symbiotic fixation, and because it had to be deliberately irrigated rather than wildly flooded, alfalfa outyielded the wild hay three to four times. Wild hay was usually cut but once a year, while alfalfa might be cut as many as three times.[17]

Alfalfa is probably native to central Asia. This legume has long been associated with developed agriculture. It was spread in the western hemisphere by Spanish colonists. Authors disagree, but sometime between 1851 and 1854 a party of goldseekers on the way to California around the Cape Horn of South America stopped in Chile and saw alfalfa being cultivated. No one knows if this party realized Chile and California enjoyed a similar climate, or if it was a matter of chance, but alfalfa seed was imported to California.[18]

George Stewart quotes E.J. Wickson, a former director of the California Agricultural Experiment Station, as indicating that W.E. Cameron had a field of alfalfa under cultivation near Marysville, California in 1851. Stewart also indicated that Mormon emigrants on the way to Utah by way of California were responsible for introducing alfalfa to the Intermountain area. The innovative agriculture genius, Henry Miller, is credited with being the first individual to widely cultivate alfalfa.[19]

Myron Angel considered alfalfa production to have started in Nevada as early as 1863. On the Humboldt River, J.P. Petigrew planted four acres of "Chili Clover" in 1864. By 1870, alfalfa became the standard forage crop of upland irrigated areas. It is important to remember that alfalfa never adapted to lowland sites along streams and never replaced native grasses in meadows. In 1879, 35,000 tons of alfalfa hay were cut from an estimated 20,000 acres in the Truckee Meadows, a sevenfold increase since 1870.[20] Fred Dangberg, the astute Carson Valley rancher, was quick to grasp the value of alfalfa and is often credited with being the first to widely cultivate the forage crop in Nevada.[21]

There are several references to the development of alfalfa fields by John Sparks for the Sparks-Harrell Company. In 1896 Sparks's ranches, the H-D ranch in Thousand Springs Creek Valley, and the Hubbard and Vineyard ranches on Salmon Falls Creek had huge fields of alfalfa.[22] One major problem with the introduction of alfalfa to northeastern Nevada was the winter hardiness of the alfalfa selections brought from Chile. Wendelin Grimm, a native of the German Grand Duchy of Baden, introduced and increased the seed supply of the first winter-hardy type of alfalfa to the United States.[23]

It took years of hard labor and expense to establish fields of alfalfa. It was a much different proposition from flooding native meadow land. Grubbing of sagebrush cost $2.00 to $5.00 per acre. Plowing the new land $2.50 to $4.00 per acre, plus $2.50 to $3.00 per acre for discing and additional $0.50 per acre for harrowing.[24]

Alfalfa was not adapted to the lowland meadows where the native hay was produced. Silt-loam-textured alluvial soils on fans and terraces were ideal sites for alfalfa production if irrigation water could be provided. The water had to be applied evenly. Alfalfa would not stand being flooded four feet deep in low places with the high spots taking care of themselves. This meant that after the sagebrush was cleared and the land plowed, some attempt at leveling had to be made. The only source of power available was horses. Crude wooden box levels were dragged across the fields after Fresno scrapers were used to move soil for major fills. Many of the alluvial fan soils were at least slightly alkaline. If small knolls were left in the field, they quickly became alkali spots as capillary rise of irrigation water deposited salts on the high spots while the rest of the field was leached.

Despite the cost, alfalfa production was highly profitable for many ranchers. In 1900, in the Lovelock Valley of Nevada, alfalfa land sold for $36.00 per acre and could be paid for in five years. Yields of alfalfa hay were five to six tons per acre. Production costs were estimated at $1.00 per ton. In 1899, 40,000 tons of alfalfa hay were fed to steers before they were shipped to the San Francisco market.[25] The raising of quality alfalfa hay and using it to finish steers before they were marketed was used as a method of modifying the environmental constraints of the sagebrush/grasslands to meet the changing demands of meat consumers. John Sparks could use registered Hereford bulls to upgrade his Longhorn cows, wet meadows to put flesh on two- to three-year-old steers, and then finish them on alfalfa hay before shipping to California markets.

For seventy-five years economists have pointed out that native meadows produced one ton per acre of hay while alfalfa fields averaged at least three tons per acre of much higher quality hay. They uniformly considered native hay production a waste of valuable irrigation water. Social activists applauded such observations as means of breaking the "big rancher" hold on the economy of much of the Great Basin.[26] There were and still are two problems with this. First, the heavy sod of the native hay lands was extremely difficult to till. The undulating meadows required considerable cutting and filling to obtain even water distribution and often drainage had to be provided. The second problem was what to do with the production of the meadow hay lands if they were broken into small farms. The climate of much of the sagebrush/grasslands limits agricultural production to forage crops that are of value only when marketed through livestock. Livestock production in this environment requires rangeland as well as farms. How to integrate extensive rangeland livestock production with intensive farming has been and still is a critical problem.

Slightly over 100 years ago, James MacDonald remarked to a western newspaper editor that cattle and grain production received about equal attention from agriculturalists in Scotland. He was amused to find the next morning's issue of the editor's paper informing the readers that one half of Scotland is pasture, and the other half is cultivated for crops.[27] It was impossible for even a literate newspaper editor to consider that the two could be an integrated part of an agricultural system. It is a basic fact of life in northern Nevada that 50 percent of the forage base for livestock production is derived from rangelands and 50 percent from irrigated lands that compose only 4 to 6 percent of the landscape. After one hundred years, the integration of peak agronomic efficiency with optimum range and animal science technology is a necessity that is no longer just amusing; it is essential.

Haying operations normally started after the 4th of July and finished by the 15th of August. In northern Nevada there was always a rush to be finished by the opening of the Elko County Fair in late August. The fair had ribbons for canned fruit and produce from the gardens of ranchers' wives. The fair featured horse racing for the more wealthy ranchers and venturesome cowboys. The fair also provided the means and the excuse for a roaring drunk by the hay hands.

The entire haying operation was conducted with a sense of urgency and rush. Irrigation water was turned out of the fields and as soon as they were more or less dry, haying began. If the sedges and wiregrass completely dried it was impossible to cut with the mower. Many ranchers raised grain on irrigated fields located on the benches. This grain, oats, or barley had to be cut and threshed after haying was completed.

The development of conserved forage as hay for winter feeding of cattle required the development of technology to make hay. The basic steps in this process were cutting a standing forage crop, drying, gathering, transporting, and storing the resulting hay. All this had to be done with men, horses, and simple mechanical implements.

Resplendent in ornate cast-iron, the horse-drawn mower started the haying process. The head mower opened up the field, cut the back swatch, and then was followed by the lesser-skilled operators in endless swaths around the initial piece. The first mower in Elko County was a wooden-framed Buckeye purchased for $225.00 by Matthew Glaser of Halleck, Nevada. At the time the cost was considered prohibitive. When the cost of mowers eventually dropped by $100 to $110, virtually every ranch in Nevada had a supply of mowers. Glaser's Buckeye mower was considered so valuable it was disassembled at the end of each haying season, the parts oiled, painted or greased, and hung in the barn to await the next rush of haying. When the reaper was first invented, independently by Cyrus McCormick and Obed Hussey, it was intended for cutting grass as well as grain; there was no distinction between reaper and mower.

Gradually two different types of machines were developed, so by 1854 there was a clear distinction between them. The cutting apparatus on the successful reaper consisted of a series of triangular section plates or blades, sharp upon the two exposed edges and fixed side by side, like large sawteeth upon the steel sickle bar. The sickle bar is given an oscillating motion as the mower moves forward and the sickle plates shear the herbage stems against the plates in the fixed guards that are rigidly attached to the bar that supports the sickle. The forward motion of the mower wheels is transferred to the horizontal oscillation of the sickle by a system of gears, a heavy flywheel, and a pitman rod. The heavy flywheel helped smooth the jerks of the traction power. The junkyard at the Walti Hot Springs Ranch in central Nevada contains the remains of eighteen mowers and this ranch had a relatively small acreage of hayland.[28]

There was nothing boring about mowing hay. The operator's attention was constantly demanded by the finger bar, the horses, and hidden irrigation ditches which gave the unsprung mowers tremendous jolts. The mower seat was a piece of sheet metal attached to a four-foot piece of springleaf steel. The resulting ride approached that of a bucking horse. The seat on the mower was positioned so the weight of the driver counterbalanced the weight of the tongue; therefore, the horses were relieved of this extra burden. Mowing was dangerous. More than one operator made the mistake of standing in front of the sickle bar only to have the team run away. Mower teams had a special affinity for taking the mower, minus a few parts, "back to the barn door." Archie Bowman, general manager of the Utah Construction Company Ranches, happened upon one of the company ranchhands picking up pieces of a brand-new mower in a hayfield. The old cowboy had started work for Sparks and Tinnin in 1884. Trying to control his temper, Bowman asked the tough ranchhand, who had spent his working life fighting against the Industrial Revolution, how in the hell had the runaway started. Just because the boss was responsible for 50,000 cattle running on 1/32 of Nevada's total land area did not matter to this man. After a chew, spit, and distasteful squint, he offered, "I guess they started about dead even, Mr. Bowman."[29]

Besides the tendency for the teams to run away, the mower operator had his attention diverted by hordes of mosquitoes that swarmed up from the vegetation as the mower passed. Besides mosquitoes, the hayfields were infested with horse and deer flies that left painful bites and could lead to runaways when the insects excited the horses. In some fields, the horses had to be covered with sacks to protect them from the insects. It was a good thing that haying camps and bunkhouses did not include facilities for bathing because the mower operators needed all the natural repellant they could muster. The hay itself formed a natural distraction to the mower operator. On windy days, and especially if the hay was tall, the waving of the grass produced a hypnotic effect that led to either sleep or seasickness.

The drag on the cutter bar, as the hay passed through the long pointed guards to be cut by the oscillating sickle plates, was hard on the team. Some ranchers in Nevada paid the highest wages to mower operators because they teamed an unbroken horse with a gentle horse on the mowers. By the end of the summer of fighting the drag of the bar, the unbrorken horse would be fit for any type of work. Many ranchers considered horses cheap and never babied the workhorses. A green team was hooked to a mower and away they went. Care was taken to not let the team run away; for once this happened, they were often ruined. To prevent runaways, green teams were worked with a trip rope attached to their legs. If the horses started to run the rope could be pulled and the horses would fall. JIC bits were often used on new teams to help control them. These two-barred bits were damaging to the horses' mouths, but helped prevent runaways.[30] The classic literary account of Intermountain haymaking was supplied by H.L. Davis. He matched a race horse against a mustang driven by a six-fingered Indian in a mower race in the Malheur Bottoms of eastern Oregon.[31]

Rocks and sticks were the bane of mower operators. If a hard foreign object in the hay passed through the fingers of the guards, it was likely to jam the sickle bar, or worse, knock out a section plate. Jamming required the mower operator to dismount and risk a runaway to free the sickle. In fields where there were a lot of rattlesnakes, it was dangerous to dismount and fumble around in the hay.

Mower operators would complete maintenance on their mowers after supper. The sickles themselves required careful sharpening on a grinding wheel. The wheel was pedal powered and the face of each section plate had to be carefully ground. The shower of sparks accompanying the shriek of grinding steel section plates was a characteristic part of warm summer evenings during haying time. This was such an important job that large ranches had a full-time sickle sharpener.

Major repairs were done by the blacksmith, but the mower operator was responsible for oiling and adjusting the machine and changing sickles. Each brand of mower had a built-in toolbox. The box contained, besides an oil can, spare guards, section plates and rivets, a cast wrench. The manufacturers of mowers must have competed to see who could design the most ornate and complicated wrench. It would have one fitting for the nut that held the pitman rod to the head of the sickle bar, another to tighten guard bolts, a third for wheel nuts, etc. In a pinch the wrench could drive staples in the gatepost knocked down while driving into the field and used to club a rattlesnake stuck in the sickle bar. An 1870 owner's manual for a Meadow King Mower indicated that year's model was furnished with two knives (sickles), two extra guards, two extra sections with rivets, a screw-wrench, an oil can, neck yoke, and doubletree.[32] The oil cans that were furnished with these early mowers were often handcrafted from copper or brass and are now collector's items.

The strain of dragging the sickle bar was hard on a team and it was customary for the lead mower to stop every few rounds to allow the horses to rest. Long after mowers were mounted on tractors, it was not unusual to see some weather-beaten old-timer sitting on his mower with the engine idling while he stared off into space, resting his horses.

To start on a 100-acre field of native hay with a 5-foot sickle bar was the height of impertinence. It took three-and-one-half hours for each mower to drop an acre of hay. If a rancher was going to winter 1,000 cows, he needed at least 1,000 acres of native hay. This works out to 3,500 hours of mowing or 350 ten-hour days. To complete haying in a month, thirteen mowers had to be in operation six days a week.[33]

After the mowed hay had wilted, it was gathered into windrows using a dump rake. Dump or sulky rakes were drawn by a two-horse team. The rake operator tripped a lever with his foot to raise the rake tongs and dump the hay in windrows. If the hay crop was very heavy, this required a strong leg. The dump rakers could always be identified in the bunkhouse because their legs twitched all night long. Hay could be bunched directly from the mower swath with a buckrake. A windrow attachment had to be carried on the sickle bar for this to work.

Raking required less skill than mowing and often was the job of boys who had gained some experience driving derrick horses. The sulky rakes were pulled by lighter horses at a trot. The rough surfaces of the native hay meadow made this a rough and dangerous ride for the young operators. J.A. Young can vividly remember the ashen faces of the hired men when they returned from finding the neighbor's son's body in the hayfield. The rake team had bolted and he had evidently bounced off the sulky at the right time to go under the dropping rake tongs.

After cutting and raking there were alternative methods available for completing the haymaking. The method followed had much to do with quality of the product obtained. If high-quality alfalfa hay was being prepared, it was gathered into shocks.[34] The hay was gathered into shocks by hand. This was the largest labor requirement of the haying process. Crews of twenty or thirty men shocking hay were a common sight. This was a low-skill job, with the only requirement being strong hands and arms to grip a pitchfork. The purpose of shocking was to reduce the amount of hay exposed to the sunlight while the drying process continued. Sunlight destroyed the green color of the forage and also the carotene pigments that serve as precursors for vitamin A. Perhaps ranchers in the 1890s did not know about vitamin A, but they did know cows did best when they ate hay with a good green color and that special aroma that goes with correctly cured hay.

When making good quality alfalfa hay, it was important to shock as soon as possible after cutting. On well-drained land, it was not uncommon to see two mowers at work in a field, followed only a few swaths behind by a two-horse rake; meanwhile, several men were making shocks as rapidly as possible so that practically all the alfalfa cut on a given day was in the shock that night. Cut at one-tenth of full bloom and properly cured, alfalfa hay contained most of the plant leaves that were important in the protein content of the hay. To get the correct drying and preservation of leaves, the timing and type of shock were important. Each shock should be the size that a pitcher could lift on the wagon with one forkful.

Slips or wagons were two methods of moving quality alfalfa hay from the field to the stackyards. In small fields where the stackyard could be reached without crossing roads and the hauling distance was short, slips could be used instead of wagons. Slips were homemade contrivances, consisting of eight 1 x 12 inch boards, 16 feet long, nailed together with a crosspiece at each end. A doubletree at one end provided the means of attaching the team. Hay was pitched on the slips with a great savings in labor compared to high wagons. The slips simply slid over the hay stubble following the team to the stack. Sometimes small iron wheels were added with a pipe axle in the middle of the slip so it would pull easier.

Hay wagons were a much more common method of transporting hay from the field to the stack. Hay wagons were sixteen to twenty feet long with a smooth bed three to four feet off the ground. At each rear corner there was a diagonally braced post four to six feet high to help hold the load. A high seat was mounted on the wagon front where it could be reached from the top of the load. A metal brake lever was below the right side of the seat and nailed to the back of the wagon seat was a forked stick or v-shaped board for tying the reins. This was the domain of the teamster.

Teamsters and pitchers worked together, usually with two pitchers throwing hay up to the teamsters who arranged the load. Teamsters on hay wagons were a semiskilled lot, while pitchers only had to be, in rancher vernacular, "hell for stout." This often resulted in a running warfare between the young, practical-joker pitchers and the older, salty, humorless teamsters. No matter where the pitcher threw up his shock it was wrong as far as the teamster was concerned. Teamsters took a very dim view of foreign objects such as rattlesnakes or skunks coming up with the hayshock. A little humor helped to make the long, hot days of backbreaking labor bearable.

If the product being prepared was low-quality grass hay instead of alfalfa, a different procedure was followed. The lowest labor requirement and, therefore, the cheapest method of transporting hay to the stack was buckrakes or sweeprakes. Buckrakes consisted of several long, wooden teeth lying almost flat on the ground, pointed at one end and fastened to a strong framework on the other. In Elko County the teeth were often made from aspen trunks. A teamster drove the rake with teeth extended down a windrow or gathered swath until hay was piled against the framework on top of the long teeth. The teamster raised the teeth slightly off the ground with a ratcheted lever and drove the load to the stackyard. Ordinarily, the rakes were twelve feet wide. With four-horse teams, larger rakes were used. Buckrakes required highly trained horses, for not only did the buckrake horse have to go ahead and stop, they had to back up. After green horses were taught the rudiments of manners by pulling mowers, they were transferred to buckrakes and the mower driver started on a new team. The buckrake evolved in the hayfields of the Great Basin. Early versions lacked a seat and the driver walked behind the team. Eventually a modified frame evolved with a single rotating wheel in the back and a seat for the teamster. Buckrake teamsters were skilled operators and commanded a better-than-average wage in the hayfields. An unskilled operator often drove the wooden teeth into the rough surface of the hay meadows or failed to pick up the hay.[35]

Hay that was handled with buckrakes became a tangled mass. The rolling action of the leading knocked the leaves from alfalfa hay. With careful attention to timing and a minimum of rolling of the hay during loading, high-quality hay could be produced with a buckrake. However, this implement was characteristically the tool used for transporting low-quality or wild hay. During the ten-hour day, one man could haul with a wagon the hay from two acres; with a slip, three acres; and with a buckrake, four acres. The buckrake was the implement characteristic of the wild-hay meadows of the sagebrush/grasslands.[36]

The low annual precipitation in most of the valleys of the Intermountain area, and snow, instead of rain, in the winter made it possible to stack hay outdoors. This was fortunate, considering the scarcity of lumber for barns in a cold-desert environment. Spoilage occurred even under semiarid conditions so the basic problem was to stack the hay as high as possible and to thatch the top well to minimize the loss. The height of the haystack became a measure of technological advancement. Horses supplied the power and various types of slides and derricks the mechanical lift.

Homemade derrick stackers came in a great variety of designs. The most numerous derricks were the "Mormon" type. Their distinctive characteristic was that the boom was pivoted upon the top of the mast. The second most popular type was the mast-and-boom derrick on which the boom extends from the side of the mast. These two accounted for 90 percent of the derricks with cable stackers and tripods accounting for the rest.

The hay was held during lifting by a Jackson Fork, a type of sling, or more rarely, harpoon forks. The Jackson Fork consisted of a triangular hardwood frame with four long, curved metal fork tongs. In an open position, the tongs were forced down into the hayload, and the frame snapped shut with a metal catch where the dumping rope attached. Once the teamster on the wagon had the fork forced into load and the catch snapped he shouted, "take-her-away." The derrick boy walked his horse a prescribed distance pulling the cable through a system of pulleys on the derrick frame which lifted the fork and hay. The derrick arm swung to the stack where the stackers shouted "dump-her" and the teamster tripped the dump rope, opening the fork, dropping the hay. The derrick boy unhooked and returned the horse for another load. The teamster pulled the boom and fork back with the dump rope. Due to the general shouting of orders with tobacco-filled cheeks, and the short attention span of derrick boys, more than one teamster-loader lost a finger in a Jackson Fork catch due to an early start.

Nets for lifting hay were of various dimensions depending on how the hay was transported to the stack. If the hay came to the stack on wagons, usually two nets were used per wagon. These nets were approximately nine by ten feet so two fit on an eighteen-foot wagon. The supports for the net were constructed of wooden poles or occasionally pipe. The net was made of ¾- to 1-inch diameter manila rope. The support poles on each end of the net had rings where the cable from the derrick attached. Down the center of the net were hooks that could be tripped by the trip rope to dump the load.

When lifting off hay wagons, one method of rigging that was commonly used was to attach one end of the derrick cable to the tip of the boom. The cable looped down to the wagon and backed up to the boom where it fed through a pully to the derrick, and then by a system of pulleys to the derrick horses. Within the loop that dropped down from the boom were two special pulleys with hooks attached. Those pulleys had special flanges that prevented them from fouling each other. The hooked pulleys were attached to the two sides of the net by the rings and as the cable tightened it pulled the net closed as it was lifted.

When hay was spotted at the stack with a buckrake, a larger net of the same basic design was used. The net was wide as the buckrake. The net tender pulled the boom back from the stack by the trip rope and the derrick boy backed the horses to lower the net down the side of the stack. When the net was near the ground the derrick boy stopped and the net tender unhooked the center of the net. The net was then lowered to the ground and the net tender pulled the outside half of the net away from the stack and hooked the ring over a pin driven into the ground. The net pin held the net in place while the buck of hay was spotted on the net. When the buckrake backed off, the net tender flipped the lifting cable over the pile and unhooked the ring from the pin and the tightened cable pulled the net signal as the boom swung across the stack and immediately began pulling the boom and net back. It was important that the net tender reversed the momentum of the swinging boom and immediately began pulling the boom and net back. More than one highly paid stacker was knocked off a high stack or killed outright by the boom and net swinging out of control.

George Stewart considered the ideal hay crew for a single derrick to be two teams with drivers to build the loads, two pitchers who remained in the field, one man to work the Jackson Fork, two men to build the stack, and one boy to drive the derrick horse. One wagon would stand at the stack being unloaded without a driver. He calculated labor costs with this crew as 3.2 man hours and 2.1 horse hours per ton of hay stacked.[37]

Cable stackers were used in areas where enormous stacks were desired. A steel-wire cable anchored securely at the two ends was supported by two pairs of heavy poles. The cable served to support a small trolley that carried the Jackson Fork or sling over the stack. Because of their inherent strength, slings were often used with cable stackers. The rope-and-pole sling was spread on the wagon before the pitchers threw up the shocks. The entire load was lifted at one time at the stack. When slings were used on a derrick, the derrick had to be supported with guy wires. Besides the derricks, there were numerous "patented" stackers. These were manufactured by equipment firms and generally were named for the blacksmith who had patented the design. The swinging Jenkins haystacker and the overshot stackers by Dain or Jackson are typical of these designs. They required a capital investment of two or three hundred dollars and, being of eastern origin, were often not strong enough for western haying. In contrast, a blacksmith in Ontario, Oregon, made the necessary ironwork for a derrick for twenty-five dollars, and the rancher could supply and assemble the woodwork himself. By establishing a reputation for quality, this blacksmith developed a market for ten to fifteen derricks a year.

In an old hayfield in the middle of the sagebrush the first thing visible, long before the field itself is apparent, is the lonesome mast of a long-abandoned hay derrick. With the pole for the mast sticking fifty to seventy feet in the air, it provides a visible landmark. In this environment where a cowboy has to ride 100 miles to find a tree sufficiently large to provide shade, where did the ranchers get these poles?

When Utah Construction operated the ranches that John Sparks founded in northeastern Nevada, they employed a full-time pole cutter named "Sanitary Bill." His name was derived from a lifetime spent free from any kind of bath. Bill spent the summers in the Sawtooth Mountain area of south-central Idaho cutting lodgepole pines for the company. He had a list of specific parts that were needed for derricks, gates, buckrakes, and horse-breaking corrals. Equipped with an axe, broadaxe, saw, and auger he spent the summer shaping parts and the winter touring ranches assembling the various ordered items.[38]

In the wild-hay meadows of the Intermountain area, hay was often stacked with slides. A slide consisted of a huge, strongly built inclined plane. The hay was brought to the bottom of the slide by a four-horse buckrake and deposited on a chain net. A cable was fastened to the net and stretched over the top of the plane and the entire stack. The other end of the cable was attached to the foretruck of a wagon, to which was hitched a four-horse team. When the load was drawn up and discharged at the proper place on the stack, the net was drawn back to the base of the plane by a single horse, adjusted, and reloaded. The four-horse buckload averaged about one ton of native hay and a load could be run upon the stack every six to eight minutes when everything was in good working order.[39]

The Beaverslide was a very popular and more complex type of slide. The Beaverslide operation consisted of gathering a load of hay from the field with a buckrake, which would then place each bunch on the hayforks at the slide base. The forks, which resembled the fork on the front of the buckrake, would carry the hay back up to the top of the Beaverslide, where the load would fall from the end of the slide onto the stack. A team of horses hitched to the cable and a pulley system was used to draw the loaded fork up the slide. As the stack was built to the top of the end of the slide, it was "topped" by the stacker and the Beaverslide was then moved forward about fifteen feet and a new bent was started. In Elko County some very large Beaverslides were built, which were capable of lifting four bucks of hay at one time. Stacks fifty feet high could be made with these very large slides.[40]

A decided disadvantage of the slide methods of stacking, especially the simple incline, was the difficulty in making the stack waterproof. With a ton of hay dropping every eight minutes, the stacker had little time to rearrange the hay to prevent loose holes that let in moisture. The height of the stack was limited to the height of the slides; generally ten to a maximum of fifteen feet. The result was a long, low, loose stack that resulted in maximum spoilage of the hay. In contrast, Griffiths describes alfalfa hay stacks built with a derrick near Ontario, Oregon, that were 375 feet long, 28 feet across, and 75 feet over the top.[41]

In local areas, ranchers evolved their own particular system of handling hay. On Willow Creek, in northeastern California, one ranch used four mowers as a cutting crew. After the hay wilted, two sulky or tumble rakes bunched it into windrows. From the windrows the hay was shocked for further curing. The shocks were picked up by a two-horse buckrake and spotted at strategic locations in the fields called yards. After all the hay was yarded, it was loaded onto wagons using a low Beaverslide. Four bucks went on each wagon. The wagons were unloaded with nets. It required a team to pull the derrick cable with these big nets. The team was attached to the derrick cart by a type of doubletree called a stretcher. The cable was attached to a cart or the running gear of an old mower and the derrick boy sat on the seat to drive the team back to lower the nets to the wagon.[42]

Often the haystacker or mower (pronounced mou-er) at $3.00 per day was the most highly paid man on the crew. The mower placed or mowed the hay on the stack. Unless each fork or sling of hay that reached the stack was carefully spread and interlocked, water spoilage would ruin the hay. There was always some loss on the top of the stack due to storms and on the entire outside surface of the stack due to bleaching. Good-sized stacks kept the bleaching from becoming proportionally appreciable. Proper stacking could also limit penetration of moisture from storms.

In Big Valley, Lassen County, California, Bill Majors was noted as the most skilled stacker in the area. With a helper he could do the work of five men on the stack. He did not stay on the ranches where he stacked hay, but walked miles across the valley to and from his home on the Lookout Indian Colony each day. When breakfast was served at 6:00 A.M., he would be at the table.

Good stackers would often taunt the net or fork tender with calls of "more hay, more hay." If the net tender got overanxious and did not watch what he was doing, the trip rope might catch the wagon or buckrake, depending on which was used to transport the hay to stack. When the rope caught, the load would dump on the way up before it reached the stack.

The middle of the stack was the most important. The first several loads of hay were dropped along the middle line of the stack with the flakes of hay flattened down and thoroughly trampled. The middle depth was three to four feet higher with a uniform slope toward the edges. After the stack was started, the outside of the stack was considered the most important part. Trampling was done so the hay was progressively looser from the middle toward the edges. Once the hay was in the stack, a crew was kept busy rebuilding the stackyard fences. This had to be completed before cattle were turned in to graze on the crop aftermath.

The dependence on hay for a forage resource for wintering cattle challenged the managerial ability of ranchers in terms of directing a small army of hay hands, raising an irrigated hay crop, and in obtaining the necessary capital to finance such operations. There were many opportunities to increase efficiency and the quality of the product produced. In the 1870s Jasper Harrell turned cattle loose on the pristine sagebrush/grasslands and marketed the increase with a minimum interference with nature. By 1900 John Sparks was faced with complexities of irrigated agriculture and making hay. Some of the Indians on Sparks's hay crews probably had existed in the sagebrush/grasslands environment as hunters/gatherers. Time in the exploitation of environments is a fickle thing with technology accelerating its passage. Haying was born out of necessity and persisted as a labor-intensive, horse-powered enterprise for fifty years in the sagebrush/grasslands.

Notes

1. Quote attributed to a disgruntled cowboy is from E.S. Osgood, *The Day of the Cattleman*, (Chicago, Illinois: The University of Chicago Press, 1929).

2. Statistics on "male" Nevada are from the burning social commentary of Anne Martin, "Nevada, Beautiful Desert of Buried Hopes," *Nation*, CXV (26 July 1922): 89–92.

3. Nevada politics is outside the limits of this volume. For a look at the different politics of Nevada, see G.L. Ostrander, *Nevada: The Great Rotten Borough, 1859–1864*, (New York: Alfred A. Knopf, 1966).

4. H.L. Davis, *Winds of Morning*, (New York: William Morrow Co., 1952).

5. The dirty plate rules are given by E.F. Treadwell, in *The Cattle King*, (New York: The MacMillan Co., 1931).

6. J.K. Woodfield, "The Initiation of the Range Cattle Industry of Utah," M.S. thesis, University of Utah, Salt Lake City, Utah, 1957.

7. *Elko Weekly Independent*, 17 April 1887.

8. J.M. Townley, "Reclamation in Nevada, 1850–1904," Ph.D. diss., University of Nevada, Reno, Nevada, 1976.

9. The Alvaro Evans papers are on file at Nevada Historical Society, Reno, Nevada.

10. Townley, "Reclamation in Nevada."

11. *Reno Crescent*, 19 February 1874.

12. *Reno Crescent*, 16 September 1871, and *Nevada State Journal*, 24 May 1873.

13. Townley, "Reclamation in Nevada."

14. David Griffiths, *Forage Conditions on the Northern Border of the Great Basin*, (Washington, D.C.: GPO, Bureau of Plant Industry, Bull. No. 15, U. S. Dept. of Agric., 1902).

15. Professor F. Lamson-Scriber's comments are cited in *Range Plant Handbook*, (Washington, D.C.: GPO, U. S. Forest Service, U. S. Dept. of Agric., 1937).

16. Creeping wildrye often occurs on saline soils. Griffiths had trouble determining saline/alkaline from non-salty soils.

17. John McCormick, J.A. Young, and Wayne Burkhardt, "Making Hay," *Rangelands*, (1979) 1(5):203–6.

18. George Stewart, *Alfalfa Growing in the United States and Canada*, (New York: The MacMillan Co., 1926).

19. Henry Miller is credited with being the first to extensively grow alfalfa, cotton, and rice in California. See Treadwell, "Cattle King."

20. Myron Angel, ed., *History of Nevada*, reprint of 1881 edition, (Berkeley, California: Howell-North Publishing, 1958); *Humboldt Register*, 2 April 1864, p. 2; *Daily Nevada State Journal*, 25 September 1874, p. 2, and 7 May 1876, p. 2.

21. "History of Alfalfa Cultivations in Nevada," in C.A. Norcross, *Alfalfa and Sweet Clover*, (Reno, Nevada: Agric. Ext. Bull. No. 5, University of Nevada, 1916). The Fred Dangberg information is from R.G. Lillard, *Desert Challenge*, (New York: Alfred A. Knopf, 1944).

22. For references to ranch improvements by Sparks, see E.B. Patterson, L.A. Ulph, and Victor Goodwin, *Nevada's Northeast Frontier*, (Sparks, Nevada: Western Printing and Publ., 1969). In 1896, alfalfa was observed by J.A. Graves, *California Memories*, (Los Angeles, California: Times-Mirror Press, 1930).

23. Stewart, "Alfalfa Growing."

24. Samuel Fortier, *Irrigation of Alfalfa*, (Washington, D.C.: GPO, Farmers Bull. 373, U. S. Dept. of Agric., 1909).

25. C.M. Saine, "Alfalfa in Nevada's Lovelock Valley," *Sunset*, VII (June-July, 1901): 31–32.

26. For an economist's point of view, see E.O. Wooton, *The Public Domain of Nevada and Factors Affecting its Use*, (Washington, D.C.: GPO, Tech. Bull. No. 301, U. S. Dept. of Agric., 1932); for social commentary, see Anne Martin, "Beautiful Desert."

27. James MacDonald, *Food from the Far West*, (London: William P. Nimmo, 1878).

28. Information on old mower is from L.W. Mills, *A Sagebrush Saga*, (Springville, Utah: Art City Publishing Co., 1954). Disassembly of mower from personal communication with Al and Norman Glaser of Halleck, Nevada. Cost of haying implements from C.A. Brennen, assisted by C.E. Fleming, G.H. Smith, Jr., and M.R. Bruce, *Cost of Producing Hay on Nevada Ranges*, (Reno, Nevada: Bull. 129, Nevada Agric. Expt. Sta., University of Nevada, 1932). Information on reapers from A.H. Sanford, *The Story of Agriculture in the United States*, (Boston, Mass.: D.C. Heath and Co., 1916).

29. N.L. Bowman, *Only the Mountains Remain*, (Caldwell, Idaho: Private Printing, 1955).

30. Personal communication to J.A. Young from Josephine S. Barterelli, Battle Mountain, Nevada. Her father followed this practice of teaming wild and gentle work horses on mowers. Comments on handling green horses are personal communication to J.A. Young from Donna Anderson.

31. H.L. Davis, *Honey in the Horn*, (New York: William Morrow and Co., 1938). This novel which won a Pulitzer Prize for fiction contains excellent descriptions of hay camps.

32. Owner's Manual for Meadow King Mower, manufactured in 1870 by Gregg, Plyer and Co., Trumansburgh, Tempking County, New York.

33. L.W. Fluharty, and J.C. Hays, *Wild-hay Management Practices in Modoc County*, (Davis, California: University of Calif., Agric. Expt. Sta. Bull. 679, 1943).

34. In some farming areas the term "haycock" was universally used; in others, "shocks" was the only term used.

35. Personal communication to J.A. Young from Donna Anderson.

36. H.E. Selby, *Cost and Efficiency in Producing Alfalfa Hay in Oregon*, (Corvallis, Oregon: Oregon Agric. Expt. Sta., Oregon State Agricultural College, 1928).

37. Stewart, "Alfalfa Growing."

38. "Sanitary Bill" was the exact opposite of what his name implied. See Bowman, "Only the Mountains Remain."

39. David Griffiths watched and described a slide work at the Island Ranch near Burns, Oregon, in 1901. See Griffiths, "Forage Conditions."

40. Description of Beaver slide prepared by Dr. Wayne Burkhardt, Assistant Professor of Range Science, University of Nevada, Reno; description of Elko Beaver slides provided by Donna Anderson.

41. Griffiths, "Forage Conditions."

42. Personal communication to J.A. Young from Virgil Knutson, Adin, California.

25. The Spanish culture contributed many of the techniques necessary to conduct cattle ranches on the open range. The roping of calves to apply an identifying brand is one example.

26. Livestock roundups were based around chuckwagons. Note the bedrolls in the wagon and the cook's box built into the back of the wagonbed.

27. The Gudgell and Simpson stock farm in Missouri greatly contributed to the development of Herefords in the West. The boy holding the bull is William Stevenson, who later became a foreman at John Sparks's Alamo Stock Farm.

28. Because of men like Sparks and Harrell, soon there were white-faced Herefords in the sagebrush.

29. After the hard winter of 1889-90, water for irrigation to produce hay crops
became a dominant factor in ranch development. Here is an early pond at
Lovelock, Nevada.

30. Cowboys soon had to muck ditches in the springtime so the native meadows
could be flood-irrigated. Mucking ditches in the springtime became an
integral part of the Nevada ranch scene.

31. Great Basin wildrye became the first source of conserved winter forage. This coarse, tall grass cures standing and sticks out above the snow.

32. Ranchers had to keep many horses to help in the cutting, transporting, and feeding of hay.

33. Pitchers threw hay shocks to load wagons.

34. Buck rakes were used to transport meadow hay to the stack.

35. To store hay for winter, it had to be stacked. The higher and tighter the stack, the less spoilage. Here wild hay is being lifted in slings by a boom derrick with guy wires.

36. Several types of derricks were developed for stacking hay. Here a boom derrick lifts hay from wagons with a Jackson Fork.

CHAPTER
10

FROM DUGOUTS TO CATTLE EMPIRES

When Texas ranchers first moved up the plains to the Northwest, their ranch headquarters often consisted of a dugout in a sidehill. But ranch headquarters evolved into an elaborate collection of buildings, built for function, not beauty or comfort. In the 1890s, a major ranch headquarters was self-contained, since ranchers were isolated for several weeks each winter and early spring. The H-D Ranch of Sparks-Harrell on Thousand Springs Creek received a shipment of supplies each fall before being isolated by winter storms, and again before haying started.

Ranch headquarters usually included a blacksmith shop, leather and harness shop, dairy, smokehouse, root cellar, laundry, bunkhouse, ice house, cookhouse, cook's cabin, chicken house, horse barn, milk cow barns, buggy house, sheds, corrals, and a big cast-iron dinner bell.[1]

The San Jacinto Ranch headquarters, which Sparks named for the battle in which Texas troops defeated the Mexican forces and avenged the loss at the Alamo, exceeded normal standards for self-containment; it had a store. The white stone store with tall windows and huge iron door and shutters was a controlling force in northeastern Nevada. Sparks and Harrell and later owners sold supplies selectively. They influenced the development of mining and small ranches in northeastern Nevada by refusing to sell supplies to newcomers, forcing them to go to Wells, Elko, or Idaho for supplies.[2]

In all her years as a resident of San Jacinto, Nora Linjer Bowman never got to see the inside of the bachelor bunkhouse. It was not considered a fit sight for the general manager's wife. The structure dated from Sparks and Harrell days, and her description would have been interesting. If you have seen one, however, you have a pretty good idea of them all—potbellied wood stove for heat, wooden bunks, bedrolls with canvas covers on the bunks, a wooden table for card playing with benches or nail-keg chairs. Worn-out boots, stray gloves, winter sheepskin coats, spring gum boots, and half-finished braiding projects accumulated under bunks and in corners. Washing was done in tin basins at the cookhouse. A trail in back of the bunkhouse led to an outhouse. On Saturday night when there was a dance at the schoolhouse, a quick talcum-powder bath and a newer shirt sufficed.

Every rancher had to enforce some control over alcohol in the bunkhouse. Often this took a passive form. Once a ranchhand dried out from a spree in town, he was too broke and too far from town to get any more alcohol. Others were more crafty, and kept rations of sweet wine hidden in the outhouse or elsewhere around headquarters.

Bunkhouses were cold in the winter and ovens in the summer. Often they were dirty and infested with vermin. Many seasonal hay hands preferred to camp in "jungle towns" located near the ranch headquarters instead. The bunkhouse at San Jacinto was constructed of lodgepole pine and covered, originally, with a dirt roof. Dirt roofs leaked with every rain and continually dribbled fine particles of soil. To counter this, canvas was tacked on the ceiling. The false canvas ceiling provided entertainment for the crew as the movements of pack rats could be traced as they alternately slid down and scratched their way along the upper surface of the sagging canvas.[3]

On most ranches the blacksmith was also the horseshoer. The ranch blacksmith shop, which many associate with glowing forge and ringing anvil, contained the smells and artifacts associated with horses. Hooked over a pole fence or piled in a corner were hundreds of used horseshoes. These ranged from small, narrow shoes for pack mules to massive, wide shoes for workhorses. There were always oddball shoes that were forged to correct some injury to the horse's hoof. In the 1890s, a sentimental ranchhand might have an oxen shoe recovered from the California Trail nailed to the wall in honor of that time, a half-century before, when work cattle challenged workhorses as a source of power.

Ranch dogs, always abundant, loved to search the horseshoeing areas for pieces of horse's hoofs that were choice morsels. After the noon meal at the cookhouse, ranchhands gathered at the blacksmith shop to ruminate before returning to work. Dog fights over which top dog got the best horse hoof chip enlivened these breaks.

Horseshoeing equipment included pincers in assorted sizes, cutting nippers, various weight-driving hammers, and a punch for the nail holes in the shoes.[4] Roundups and haying were busy times for the blacksmith. Rank, young horses had to be tied and thrown before their shoes could be applied. Handling horses in this way was rough and dangerous. To throw and control a rough horse with ropes required skill. A good horse could be ruined or the ranchhand or blacksmith injured if the job was not done correctly.

Even broken horses could be dangerous to shoe. Some horses developed devilish schemes to make the blacksmith's life miserable. To examine a hoof, the farrier lifted the hoof and held it between his legs. By gradually increasing the weight placed on the leg the shoer was holding, the horse could bring a backbreaking load to bear without the farrier's noticing. Many horseshoers had a mean outlook on life, which probably developed from working all day bent over, supporting part of a horse while being kicked, bitten, stepped on, and generally roughed up. Many carried large rasps in their tool kits to teach manners to misbehaving horses.

If the horse had shoes when brought into the blacksmith shop, the shoer examined the old shoes and hoofs for signs of uneven or unusual wear. The old shoes were pulled with dull nippers and the hoof was cleaned with a horseshoer's knife. The hoof was trimmed and shaped with a six-row rasp. In the forge the shoe was heated to black-hot, not red-hot, and applied to the horse's hoof. The shoer noted the burns on the hoof and used his rasp to reshape the hoof until it was perfectly flat.

The shoe was then nailed to the hoof with big-headed horseshoe nails. The points of the nails emerged through the side of the hoof and were bent over. If the horse made any violent movements at this stage of shoeing, the shoer stood a good chance of having the bent-over nails stuck in his legs. Horseshoers wore heavy leather aprons to protect their legs from the nails. The bent-over nails were clinched with a sharp hammer rap while a clinch bar was held tightly against the nailhead. Sharp nippers were then used to cut off the nails and a curled face clincher used to bend the clinched nails firmly into the hoof.

In the fall, before the winter hay feeding started, the teams that were to be kept at the ranch headquarters to pull the feed wagons were shod with caulked shoes to give them traction in the winter's mud and snow. The blacksmith welded the caulks onto smooth shoes. Cans of Cherry Heat Welding Compound served as a flux when shoes and caulks were heated red-hot and welded together.

Once or twice a year, usually before haying, wagon wheels had to be reconditioned. The wheels were removed and a rod was placed through the axle hole. The wheel, suspended by the rod, was placed in a sheet metal pan of boiling linseed oil and was slowly rotated until all the wooden parts were reconditioned with absorbed oil. If the metal rim was worn or loose, the blacksmith replaced it. The finished rim was heated, placed over the wooden wheel, and shrunk to a tight fit as it cooled.

The blacksmith for John Sparks had to be especially skilled, as Sparks was a hot-rod buggy driver who reached for the whip when other drivers reached for the brake. In 1896 he was driving on the fourth set of wheels on his custom-made buggy when he gave the banker, Jackson Graves, a hair-raising ride from the H-D to the Vineyard Ranch.[5]

The hubs of all wheels, from buggies to hay wagons, were lubricated with axle grease. This hard, sticky grease came in a gallon pail which was saved and cleaned after it was empty for use in measuring. "About an axle grease bucket's worth" was common ranch English.

The blacksmith shop was the center for harness and saddle repair. A workhorse harness was largely leather, but to function, it required a bewildering variety of snaps, buckles, and chains. Harness supplies included utility cockeyes, harness snaps, dixie belt snaps, breast strap roller belt snaps, open eye snaps, and snaps and thimbles. For harness repair, a ranchhand sat on a wooden workhorse that supported a long jawed wooden vise at one end. Heavy needles, awls, waxed thread, leather punches, and knives were his tools.

The harness attached to the horse collar with a pair of curved hardwood sticks called haines. Commercial teamsters often decorated the tops of haines with bows from which bells were hung. On ranches, the haines usually ended in polished brass spheres. Just as truck drivers today polish the chrome on their rigs, teamsters took pride in polishing haines ornaments. The teamster sometimes added ornaments to the headstalls of his team's bridles even though the harness and horses belonged to the ranch. If the teamster wintered in California, the ornaments were conchos made from silver dollars or Mexican pesos by Mexican silversmiths. A more common practice was to salvage the back of a broken pocket watch and attach a wire loop to the back by filling it with lead or solder. Simple things these ornaments, but they were symbols of pride.

The jewel of every large ranch headquarters was its garden. Many ranch headquarters were at elevations between 5,000 and 6,000 feet, but a broad range of produce was obtained from the gardens. During the growing season, the garden was the only source of fresh vegetables.

Potatoes, cabbage, and root crops, especially carrots and parsnips, were stored in underground cellars for the winter. Many crops were canned in glass jars for winter at the constant risk of botulism.

In the spring, when vegetables from the cellar were half-rotten, they were a poor excuse for food, but better than nothing. A mail stagedriver in central Nevada hated the springtime. In addition to the muddy roads, he had to face invitations to lunches from ranchers—sometimes boiled, thin venison and rotten cabbage dug from the root cellar. The visit from the mail stage was important. Besides the mail and papers, he delivered news that was too trivial and too personal for the local papers.[6]

The choreman tended the garden, as well as the milk cows, chickens, and pigs. There were lots of jokes about Texas cowboys who had never tasted cow's milk. On late nineteenth-century Nevada ranches the milk cows provided milk, butter, and cottage cheese for the cookhouse, and skimmed milk for the pigs.

The late nineteenth-century cowboys ate more cured pork than he did fresh beef. The pigs were slopped with table scraps and skimmed milk. Many ranches raised grain in rotation with alfalfa, and this grain was fed to the pigs. The pigs' rations would have been deficient in quality protein without the skimmed milk.

As far as the ranchhands were concerned, the cookhouse was the focal point of ranch headquarters. It was located outside the kitchen and was simple, with long, wooden tables covered with oilcloth. The kitchen was massive, with implements large enough to feed forty men three meals a day. The kitchen had a huge black stove with tiered warming ovens and a tank water heater. For breakfast, great black hotcake griddles were put on top of the stove. These griddles produced hotcakes eight inches in diameter and one-half inch thick. It was a good thing there were few deep bodies of water in the sagebrush/grasslands. After a breakfast of hotcakes like these, swimming would have been difficult.

The cook was very important to ranch headquarters. No matter how hot the weather was, the cook still had to feed wood to the stove. Even if the men ate a cold meal, the stove had to heat water for the dishes. On small ranches the herculean task of cooking for haying crews fell on the rancher's wife. One prominent Nevada ranch lady relates a terrible nightmare in which she dreamed she was in hell cooking in midsummer on a wood stove for a forty-man haying crew. She knew she was in hell because, when the crew came to sit down, each wore the uniform of one of the prominent federal land management agencies.

Besides giant hotcakes, breakfast consisted of home-cured ham or bacon; mush, either creamed or fried; and dozens of fried eggs. At some ranches it was traditional to have beefsteak for breakfast. If the cook failed to have steaks ready for the crew, they would probably go down the road. Just as important as steaks at some ranches were baking-powder biscuits. The food was served on heavy ironstone plates. Gallons of boiled coffee were served in handleless mugs. The mugs were so thick that the scalding coffee never cooled. For those who did not care for scalded tongues and tonsils, the coffee was slurped from a saucer. The old-timers on the ranch crew treated the cook and his helper with cold disdain. It was beneath their dignity to ask the cook for a second cup of coffee; they simply tapped the empty mug with a knife handle.

The noon meal was simple, but tremendous in bulk. Boiled potatoes, fried thinly cut steaks, white or brown gravy, beans and ham hocks, and home-baked bread or biscuits with farm-churned butter. During the summer, lettuce was served wilted with vinegar and bacon-grease dressing. A few fresh vegetables went a long way. Rice or canned peaches served for dessert.

If a large haying crew was being fed, they might have beef every day. It took a large crew to consume a beef before it spoiled. When a beef animal was killed, the steaks were cut thin before frying. The grass-fed animals had little marbled fat and the meat was tough. The teeth of the older ranch hands were not up to chewing thick hunks of such meat.

Facilities for slaughtering beef were primitive at most ranches. Usually a windlass was arranged in the corral between two stout gate posts. The animal selected for slaughter was herded into the corral and shot or knocked in the head with a sledge. The animal was hoisted with the windlass and bled. The offal was dumped on the ground and left for the pigs. In the early days of ranching, during the 1870s, the offal was often given to Indian families that had become attached to the ranches when cattle outcompeted them for grass seeds. The hide was thrown over the fence, hair side down, to dry for conversion into rawhide. The meat was cut up and hung in a screen cooler. Small ranches often had cooperative exchanges of meat so it could be used before it spoiled. In the summer the only alternative was to put the excess meat in brine as corned beef.

Beans were a standard item on cookhouse menus, and cooks who left rocks in the beans were not popular with the crew. The lazy cook's method of cooking beans was to dump them in the pot without cleaning and let the chaff float to the surface and the rocks settle to the bottom. The evening meal often featured a large roast or two main meat dishes like baked ham and roast beef and more boiled potatoes and gravy. The heat in the kitchen during the evening meal and through cleanup reached boiler-room intensity. During the late summer, door and window screens were black with flies waiting for an opportunity to pop in.

After the wreckage of the evening meal was disposed of, it was time for the cook to do the next day's baking. Besides bread, the crew thrived on huge slices of fruit pie—dried apricot and raisin during the winter and apple in season. If the cook was in a good mood, he might prepare donut dough to fry for the next morning's breakfast. The kitchen started to cool late in the evening, especially if the ranch was situated at the mouth of a canyon where breezes rustled the cottonwood leaves. The cook had a few moments to sit on the porch and enjoy the evening before dropping exhausted on a bunk in a hot, stuffy room with mud-dauber nests on the peeling wallpaper. At 4:00 A.M. he was up to light the barely cooled wood stove to start another day.

Who were these supermen who ruled the cookhouse? Often the cook was a dried-up 90-pound Chinaman displaced from some Central Pacific construction crew by way of a worn-out mining claim. The cookhouse was a significant cost of production facing ranchers. The cost of boarding labor for growing hay was 29¢ per ton and the cost of boarding the haying crew was 63¢ per ton. This was with a total labor cost of $3.68 per ton of grass hay produced.[7] The cash expenditures for supplies for the cookhouse included flour at $3.50 to $4.50 per cwt, sugar $6.00 to $8.50 per cwt, beans $7.00 to $9.00 per cwt, and coffee beans at 33¢ to 50¢ per pound.

Henry Miller, the master of nineteenth-century ranch management, looked closely at the vegetable garden when he visited a ranch. He chastised one of his managers with; "Your vegetable garden is not big enough for the men you have. Good vegetables make the men more

191

content and draw a better class of laborers." He surprised another one of his managers by examining the slop bucket in the back of the cookhouse, "Those potato peelings are too thick. You can tell a good housekeeper by looking at the potato peelings."[8] The ranch manager had to balance the welfare of the crew and the cost of production. If he had the cook skimp on food, the crew went down the road.

Cooks were notorious for having touchy feelings. One cook posted this sign in the cookhouse: "If you can't wash dishes, don't eat. We use wood in the cookstove cut 16 inches long but no longer. A busy cook loves a full woodbox. A full water bucket makes a happy cook, stray men are not exempt from helping wash dishes—bring wood or water. The well is just 110 steps from the kitchen, mostly downhill both ways."[9]

By the 1890s, most of the cowboys in northern Nevada and southern Idaho rode single-cinch saddles, with Walker trees made by the Visalia Saddle Company of San Francisco. The leather used in these saddles was oak-tanned and at least seven years old when the saddles were made. Saddles and horse equipment for the sagebrush ranges of the Intermountain area underwent considerable evolution during the three decades of intensive livestock production on the sagebrush/grass-lands. Texans brought their equipment and traditions when they delivered the first cattle. The Spanish vaquero from Old California contributed a different style of equipment.[10] In the saddle room or blacksmith shop there was a chest devoted to horse medicine. Horses are prone to injuries, especially around the feet and legs. Bottles of Bone Blister, White Liniment, Badger Balm, Lucky Four Blister, Germ Killer, and Antiseptic-Poultice were part of the mystery of treating unsound horses.

The first reaction of settlers in a treeless, semiarid environment was to plant trees to modify the environment of the ranch headquarters into something that fit their conception of desirable surroundings. Very few trees are adapted to the sagebrush/grasslands. High on the sides of the fault-block mountains are found scattered groves of quaking aspens, but aspens are not easily adapted to the environs of most ranch headquarters and quaking aspens are not easy to propagate. Other members of the poplar family did adapt to conditions around ranch headquarters and were widely planted. In the western Great Basin, Fremont cottonwoods form a natural gallery forest on the banks of the

Truckee, Carson, and Walker rivers and extend well out into the Carson Desert. In the mountains of northeastern Nevada in streamside environments, the black cottonwood occurred natively. Much of the course of the Humboldt River from Elko to the Humboldt Sink was devoid of trees under pristine conditions. Settlers found it relatively easy to propagate the native cottonwoods by root cuttings or sucker sprouts with roots. Buildings, cabins, corrals, and bunkhouses were shaded by the rapidly growing cottonwoods. The Lombardy poplar was an import that was widely planted around ranches and homesteads. A relatively short-lived tree, the Lombardy poplar often succumbed to drought and disease. The naked skeletons of these trees mark innumerable abandoned ranches and homesteads.

The ranch headquarters was the only part of the sagebrush/grass-lands ranching system where women regularly played a role. There were occasional female ranchers, or even outlaws, but generally ranching was male-dominated. The life of the rancher's wife was hard and lonesome even if she lived at ranch headquarters. The distance between ranch headquarters on the sagebrush/grasslands precludes close contact with neighbors. Often the only companionship available for the ranchers' wives were the Indian women who washed clothes and worked around the ranch house.

There is a folk story in northern Nevada that tells the reaction of a Crowley Creek rancher's wife to the sagebrush environment. The Crowley Creek woman was left completely alone at her stone-walled ranch house for considerable periods while her husband supervised roundups. One year before the fall roundup she begged her husband to have a party before he left. The invited guests came for a hundred miles across the sagebrush valleys and barren playas to attend. When all the guests had assembled in the ranch yard and pitched their traveling tents beside their rigs, the Crowley Creek woman went in the house and killed herself by placing a rifle barrel in her mouth and firing. She left a note saying she did not want to die alone in this empty land.

The winter haying crews were closely identified with ranch headquarters. In contrast to huge summer crews that were strictly seasonal, the winter hay feeders were there year-round. Once the summer hay crew departed to a warmer climate, the year-round crew often ate their meals with the foreman or on smaller ranches with the owner. Hay feeding did not start until early winter. Cattle were gathered from the summer range in October and grazed on crop aftermath. This was the regrowth on the native-hay meadows.

On Sparks-Harrell ranches, the cattle were worked and sorted as they were gathered. Dry cows and steers that were not ready for market were pushed out on the salt desert vegetation of the Bonneville Salt Flats northeast of Montello or, on the northern ranches, sent to the Owyhee Desert. Cows with small calves and replacement heifers were kept on hay aftermath and, when that was exhausted, were fed hay.

If the winter was severe, weaker animals outside the fenced fields were worked into the fields and fed hay. In central Nevada during a hard winter the cows and calves were in the feedyard area and the yearlings and two-year-old steers walked the fence trying to get in. The feeding crew would warn the foreman that the steers were getting weak and should be let in for supplemental feeding. The ranch foreman finally would relent and send the men to bring in the steers. Very few steers were found alive. One of the cowboys explained that the steers wore such a deep trail around the fence, the sides caved in and the steers disappeared.[11]

Under average conditions, with stockyards and feed grounds conveniently located, one man could feed and take care of 400 head of stock cattle during the hardest winter months. Although one ton per brood cow was a general hay requirement, ranches that bordered salt desert areas might get by with one-half to three-quarters of a ton of hay per cow. In high mountain valleys such as the headwaters of the Humboldt River, Salmon Falls Creek, and Goose Creek, one-and-one-half tons of hay were required for each cow.[12]

Some ranchers treated hay in the stack like money in the bank; they fed only as a last resort and then as little as possible. Their cattle went on the range in poor condition each spring, but they had hay reserves carried over from winter to winter which were invaluable during severe winters. Ranchers who subscribed to this philosophy often fed heaviest during the early part of the winter so the stock was in good shape for the most severe weather and then greatly reduced the amount of hay fed during late winter and early spring. This made the cattle rustle for their own forage during early spring.

Grasses native to the sagebrush/grasslands grow very little during winter due to low temperatures. When it warms up in the spring, these grasses must use their carbohydrate reserves to renew growth while soil moisture is available. If native perennial grasses are grazed by the same class of livestock at the same time each spring, the grasses die. The earlier in the spring the grasses are grazed, the greater the mortality rate. Ranchers that fed heavy early in the winter and tapered off later were forcing cattle to use the perennial grasses before the grasses had a chance to rebuild their storehouses of carbohydrate reserves.

John Sparks was miserly with his hay supplies. When Jackson Graves toured the ranches in 1896, the stockyards on the Salmon Falls Ranches contained stacks that were three years old.[13] It was only six years since the disaster of 1889–90, and the memory of thousands of dying cattle was slow to dissipate.

After Thanksgiving when hayfeeding started, it was not such a bad job. Once the cows got their hay and the horses were taken care of, the winter crew worked on other chores. On stormy days the crew sat around the stove in the blacksmith shop and worked on harness and saddle repair and watched the ranch dogs steam before they grudgingly moved back from the stove. By March the novelty was gone. For seven days a week it had been the same thing every morning, snow, rain, or shine. The tempo of ranch life quickened in early spring. It was time to muck ditches, get the irrigation dam installed, and brush the cow chips on the meadows. Instead of being the focal point of the day, hay feeding was an extra chore tacked on top of everything else. The rancher was more than willing to turn the cows loose. For their part, the cows were sick and tired of a steady diet of dry hay. The lure of fresh grass in the hills drew them away from the fields.

One of the tenets of twentieth-century range science is that grazing the same plant species at the improper season with the same class of livestock will result in the disappearance of the forage resource. This has been proven time and time again with early spring grazing, but ranch management systems evolved so it was almost impossible to escape this type of use of the grazing resource.

Feeding hay contributed to overutilization of rangelands located close to the ranch headquarters. This may seem impossible, especially since it was just stressed that hay should have been fed longer in the spring in order to give the native grasses a chance to build up carbohydrate reserves before the cattle were turned out on the range. The opposing view was expressed many years ago by Lewis "Broadhorns"

195

Bradley, a nineteenth-century governor of Nevada. "There is a good deal in educating a critter. He is like a man—if he knows his living will depend on his rustling, he'll rustle!"[14] The winter of 1889–90 proved that some winters occurred when hay feeding was absolutely necessary for survival of the animals. However, feeding hay kept the stock concentrated near the feed grounds, which resulted in severe overutilization of native perennial grasses at the worst possible season.

The nineteenth-century ranch headquarters grew into a self-contained micro-community. The ranch owner accumulated a considerable capital investment in buildings, tools, and equipment. Board for the resident ranchhand became a major expense. The concentration of man's activities and the grazing animals around the feed yards made the ranch headquarters the most degraded portion of the environment of the sagebrush/grasslands.

Notes

1. The description of a typical nineteenth century ranch headquarters is from a description given by Mrs. Agnes Robison about the McGill Ranch in White Pine County, Nevada. See also Orville Holderman, "Jewett W. Adams and W.M. McGill: Their Lives and Ranching Empire," M.S. thesis, University of Nevada, Reno, Nevada, 1963.

2. F.C. Scharader, *A Reconnaissance of the Jarbidge, Contact, and Elko Mountain Mining Districts, Elko County, Nevada*, (Washington, D.C.: GPO, Bull, 497, Geological Survey, U.S. Dept. of Interior, 1912).

3. M.F. Knudtsen, *Here is Our Valley*, Helen Marye Thomas Memorial Series No. 1, (Reno, Nevada: Agric. Exp. Sta., Max C. Fleischmann College of Agric., University of Nevada, 1975).

4. Personal communication to J.A. Young from old-time horseshoer Bill Smith, Etna, California.

5. J.A. Graves, *California Memories*, (Los Angeles, California: Times-Mirror Press, 1930). Years later this custom-made buggy was found in back of a store in Montello and restored by the Bowman family.

6. Knudtsen, "Here is Our Valley." Sparks at one time had a contract with the U. S. Government to deliver mail from Montello up Thousand Springs Valley and down Salmon Falls Creek into Idaho. Essentially, the government paid the expenses of shipping supplies to the Sparks-Harrell ranches.

7. C.A. Brennen, assisted by C.E. Fleming, Grant H. Smith, and M.R. Bruce, *Cost of Producing Hay on Nevada Ranches*, (Reno, Nevada: Bull. 124, Nevada Agric. Expt. Sta., University of Nevada, 1932).

8. E.F. Treadwell, *The Cattle King*, (New York: The MacMillan Co., 1931).

9. W.M. Raine, and W.C. Barnes, *Cattle*, (New York: Grosset and Dunlap, 1930).

10. Adelaide Hawes, *The Valley of Tall Grass*, (Bruneau, Idaho: Private Printing, 1950).

11. Knudtsen, "Here is Our Valley."

12. Brennen, et al., "Cost of Hay."

13. Graves, "California Memories."

14. L.W. Mills, *A Sagebrush Saga*, (Springville, Utah: Art City Publishing Co., 1954).

CHAPTER
11

HEREFORDS IN THE SAGEBRUSH

The bulk of American cattle during the mid-nineteenth century, especially those that belonged to farmers of the frontier districts, were of poor quality. The Longhorns of Texas were virtually a free-breeding population with no conscious effort toward quality improvement by herdsmen. The milch cows that followed the covered wagons to Oregon or California were generally plain grade animals of poor quality. By the early nineteenth century, the concept of improved livestock breeds was in full blossom in England and Scotland. As the roots of the Texas longhorn reach back to the open ranges of the Extremadura in Spain, the English cattle breeds in the sagebrush/-grasslands extend back to the countries of the United Kingdom.

In colonial times, country gentlemen of the largely agricultural society followed the image of English or northern European landowners to make their estates showplaces of agricultural technology. Status was provided by importation of an English bull or a Belgian stallion. Eventual transfer of this concept to the frontier was through the local fair and livestock show. After the Civil War, the railroad transportation system had sufficiently evolved in the upper Midwest and eastern portion of the country to allow statewide and regional livestock shows. The increased public exposure heightened interest in improved breeds of cattle and the livestock industry responded quickly. Nearly every rancher could see the Longhorn had many drawbacks as a red-meat producer. There was no practical way to replace the millions of Longhorn cows used in stocking the new rangelands, but herds could be gradually improved by removing Longhorn bulls and replacing them with quality English breed sires. With this would come a huge market for bulls.

Improved breeds did not win instant or widespread acceptance in the West. When the Scottish agriculturist, James MacDonald, visited Texas in the 1870s, he doubted that Shorthorns would ever be a success in Texas, because few of those that had been introduced had survived more than twelve months.[1]

The fertile Tees Valley, in northern England, was the cradle of the

Shorthorn. In this section, late in the eighteenth century, there was born a breed of cattle destined to become the most numerous improved strain during the nineteenth century and one of the most valuable of the world's breed.[2] The counties of York and Durham, in the Tees Valley, had long been known as the home of a type of cattle known as the Teeswater breed. They were characterized by size of frame, strength of bone, and ability to attain great weights at maturity. In color, these ancestors of the Shorthorn were of various combinations of light yellowish-red and white.

Throughout most of the eighteenth century, breeders of the Teeswater stock were generally in accord as to the type of cattle best suited to their operations. It was not, however, until within the last two decades of the century that the breed began to take definite form. Molding of the Shorthorn was initiated by the skillful hands of two brothers, Charles and Robert Colling, tenant farmers in Durham. Their foundation material was secured in 1783, and was comprised of the bull, Hubback, and a number of cows purchased at the Darlington market.

Charles Colling was a student of Robert Bakewell, the first man who bred livestock scientifically. Quick to discern the potential value of Teeswater cattle, Charles and his brother set out to improve and establish the type. Their first ten years were spent experimenting with breeding material approaching their ideal. The Collings finally produced a bull named Favorite that met their requirements. They immediately applied Bakewell's mating system of inbreeding. They mated Favorite to his daughters, granddaughters, and occasionally to his female descendants in the fourth and fifth generations. Comet, an illustrious sire, the first Shorthorn to sell for $5,000, came from the mating of Favorite and young Phoenix, a heifer that had been produced from the union of Favorite with his own dam.[3] To the unfamiliar, this inbreeding may be shocking, because of the increase in expression of recessive characteristics. Remember, however, that inbreeding also increases the homozygosity of desirable as well as undesirable genes. If the breeder can withstand the losses associated with rapid inbreeding, and make desirable selection while inbreeding, he ends up with the desirable traits he selected fixed in his herd. These desirable characteristics will be passed on to the offspring of the herd because the population is homozygous. The Tees Valley cattle were a variable population that represented a portion of the greater population that made up the species. When the Colling brothers finished their inbreeding program, they had further circumcised the genetic variability to form the basis for the breed called Shorthorns.[4]

The intensification of the bloodlines Favorite imparted to the Colling Shorthorns showed an improvement in fleshing ability and a refinement in quality that soon drew the attention of the contemporary breeders. In a shrewd piece of advertising, the Colling brothers fitted and showed throughout the United Kingdom a celebrated "White Heifer that Traveled" and the 3,400-pound Durham ox. This focused the attention of the British public upon the new breed.

During the 1830s the center of Shorthorn development shifted from England to Scotland. Amos Cruickshank, of Sitlyton, Aberdenshire, developed his Champion of England bloodlines. The Scottish lines of Shorthorns were a vigorous, early-maturing, and well-fleshed breed that immediately became popular.[5]

The first Shorthorns to reach America were brought to Virginia in 1783 by the firm of Gough and Miller. By 1800, cattle bred from this importation penetrated to Ohio and Kentucky. These early importations were of English or Colling-type shorthorns. After 1880 the Scottish type of Shorthorn rapidly increased in popularity in America. The Scottish type fit the need of the times for a vigorous, early-maturing animal that did not require a huge quantity of feed or a long feeding period to be finished for slaughter.[6] Show-rings of the major livestock exhibitions became the battlegrounds for the purebred cattle industry in America. The overweight, fitted, and pampered animals of the livestock shows became the standards of the industry. These animals would probably not have survived if turned loose on the western range. The breeders were not interested in selling the show animals, except for an occasional herd bull to another breeder. They were aiming for carloads of bulls being shipped to the western range at handsome prices. The glitter of the show-ring and the overweight, fitted animals were illusions created for the western rancher who journeyed east with cash in hand.

Improvement of the Spanish-type longhorn began in Nevada in the 1870s. In 1874, Senator Gabriel Cohn and John M. Dorsey of Lamoille in Elko County, Nevada, purchased from Saxes of Kentucky recently imported Shorthorn bulls with a value of $5,000 to $20,000.[7] Their offspring were worth seven dollars to ten dollars per head more than straight Longhorn steers. Also, the heifers that resulted from mating the Shorthorn bulls and Longhorn cows proved to be excellent brood cows. The hybrid heifers were fertile, vigorous, good milkers, and produced larger, vigorous calves. In summation, they expressed all the qualities associated with hybrid vigor.

There is much confusion in nineteenth-century accounts over the names Durham and Shorthorn. Many people considered Durhams to be a distinct breed. The Durhams were all-red animals while the total Shorthorn breed included red, white, and roan animals. James MacDonald was shocked to find that Americans preferred all-red Shorthorns in the 1870s, and that Americans were paying a premium for all-red animals. He was eager to return to Scotland to export red animals to America for superior prices.[8] The origin of the name Durham has been linked with the County Durham in the Tees Valley. American importers of Shorthorns apparently attached considerable prestige to this geographical location.[9]

Shorthorns were not the only English breed imported to upgrade the Longhorns. Galloways were running on the North Fork of the Humboldt. The English-financed Nevada Land and Cattle Company imported polled-back Angus bulls from Aberdeen, Scotland. For some reason this affronted the Spanish cowboys on the ranch who promptly roped and castrated the expensive bulls.[10]

Shorthorns were the most numerous English breed of cattle on the nineteenth-century western range although many other minor breeds were represented. By the turn of the century a second major breed suddenly became popular and soared in numbers to rank second behind Shorthorns. This breed, the Hereford, was predominantly red on the body with white markings, and with a trademark, a white face. Herefords were destined to become extremely popular in sagebrush/grasslands range. Although among the early breeds introduced, Herefords were little-known in the United States until the opening of the range country. Beginning about 1880, Herefords made rapid growth until they were second in numbers to Shorthorns and almost supreme on the range.[11]

The earliest record of Hereford cattle came from the Wye River Valley in the country of Herefordshire. For centuries, Herefordshire had been noted for good grass and excellent beef. There is considerable evidence that the Hereford breed antedated the systematic selection and breeding of other pure breeds of cattle in Great Britain. Benjamin Tompkins and William Galliers, Sr. were breeding cattle of the Hereford type before Robert Bakewell began his historic improvement of the ancient English longhorns. Benjamin Tompkins and others apparently possessed herds of a well-established type of Herefordshire cattle in the late eighteenth century, when Charles and Robert Colling were barely laying the foundation of the modern Shorthorn.[12]

The origin of the white face of the Hereford is shrouded in mystery. Some traditions credit the marking to Dutch cattle; others attribute it to cattle brought from Yorkshire. The white face was a fixed characteristic in many herds as early as 1788, although it was not until many years later that it became the universal standard for the breed. Pedigree registration for Herefords was not on a firm basis until 1878 when the English Herd Book Society was founded. In the years preceding this move, there had been much difference of opinion as to color of the breed, and the first two volumes of the herd book recorded mottle-faces, white faces, and grays.[13]

The first importation of Hereford cattle to America for which there is definite records was made in 1817 by Henry Clay of Kentucky. In 1840, the first aggressive efforts to establish the breed in America were undertaken by William H. Sotham, who in a partnership with Erastus Corning of New York, imported twenty-two head. The Hereford became a significant breed in the United States in the 1870s when T.L. Miller of Beecher, Illinois, and Thomas Clark of Ohio began importing cattle. The Miller and Clark herds attracted more attention than any of their predecessors and a number of men were drawn to the breed shortly before 1880. Among the herds founded at this time were those of C.M. Culbertson of Illinois, Fowler and Van Natta and Earl and Stuart of Indiana, and Gudgell and Simpson of Missouri.

Of the American foundation herds that influenced the development of the breed, none was more important than that of C.J. Gudgell and T.A. "Governor" Simpson of Pleasant Hill, Missouri. Gudgell was a banker and Simpson was a farmer and horse dealer who formed a partnership to raise Shorthorns. In 1876 Gudgell and Simpson visited the Centennial Exposition in Philadelphia where they saw Herefords. They liked what they saw, except that Herefords generally were light in the hindquarters. Later they saw a bull calf imported from England by C.M. Culbertson of Newman, Illinois. This bull, Anxiety 2238, had the hindquarters that Gudgell and Simpson believed the breed required. Unfortunately, Anxiety 2238 died before it had a chance to influence the developing American strain of Herefords. In their search for a foundation sire, Simpson went to England to find a relative of Anxiety 2238. As he prepared to leave, his partner gave Simpson this parting advice, "If you find a bull with an end on him, bring him with you." When he returned with Anxiety 4th, he was known as "the bull with an end."[14] Anxiety 4th became a great herd bull and helped establish Gudgell's and Simpson's operations as one of the top Hereford herds. Anxiety 4th died in 1890, but his offspring contributed to Hereford herds throughout the Midwest and in increasing numbers on the western range.

Not only were Hereford bulls popular on western rangelands, but herds of purebred Herefords soon became established on the western range. George T. Morgan, an Englishman, came to Wyoming in 1876 with the idea of introducing breeding stock. He returned in 1878 with a shipment of Hereford bulls consigned to A.H. Swan. These bulls, reportedly the first in Wyoming, cost over $10,000 to import. Swan, with Morgan as manager, established the Wyoming Hereford Ranch on 40,000 fenced acres. Through the mid-1880s, Wyoming Hereford Ranch imported over 500 Herefords from England.[15]

Joseph Scott of the 71 Ranch at Halleck is usually credited as the first to import Herefords to Nevada. He imported twenty heifers and four bulls in the late 1870s. Scott, born in 1884 in Ireland of Scottish parentage, developed ranches in Montana, eastern Washington, and Wyoming before coming to Nevada. He came to prominence in the purebred Hereford business in the firm of Scott and Hank of Mandel, Wyoming. They bought foundation stock from C.M. Culbertson, and also imported stock from England.[16] In 1880 Scott and Hank were taxed on the Elko County tax rolls for fourteen head of purebred Herefords valued at $100 each. By 1886 they had sufficient purebred stock to sell a yearling to Russell and Bradley for $400 and a cow and calf to Williams and Hunter for $1,000.[17]

The purchase of quality bulls to improve their herds involved considerable capital investment for ranchers. Another early Nevada leader in the importing of quality bulls was Abner Cleveland, cousin of the president.[18] The prominent rancher and later governor, Jewett Adams, used a unique method of raising capital for a carload of whiteface bulls. He won $30,000 in a high-stakes poker game and promptly reinvested in Herefords.[19]

In 1885 the Elko newspaper announced that John Sparks of the giant ranching firm of Sparks-Tinnin was leaving Elko County, where he had resided since 1881, for the western Nevada city of Reno. The reason for this move was given as improving educational opportunities for his children. Sparks had been living at the relatively remote H-D Ranch on Thousand Springs Creek. He had credited the hot springs located on the H-D Ranch with restoring his wife's health during that period. Despite the medicinal waters of the H-D hot springs, John Sparks left Elko County for the Reno area.[20] Sparks continued to operate the extensive ranching properties in northeastern Nevada and south-central Idaho after he moved to Reno. He also maintained a residence in Georgetown, Texas.

In 1887 Sparks made another significant change. He purchased property south of Reno known as Anderson's Station, located on the former site of the Junction House. This was a way station which lay in the center of the major immigrant trails passing through the Truckee Meadows area.[21] Sparks, with a flair for the colorful, named his new property Alamo for a nearby grove of cottonwoods. The property contained about 1,640 acres. Sparks constructed a striking plantation-style residence. The Alamo Stock Farm was located in the geothermal belt north of Steamboat Hot Springs and two artesian wells were located on the property. One well tapped hot water at 240 feet below the surface and the second penetrated to a depth of 560 feet. The hot water was funneled into a large swimming pool.[22] Perhaps Mrs. Sparks may not have missed the hot springs of the H-D Ranch.

Later Sparks purchased the 400-acre Mayberry Ranch located on the Truckee River west of Reno. This property increased the irrigated pasture and hay land under Sparks's control in the Truckee Meadows area. The Virginia City and Truckee (V & T) Railroad tracks bisected the Alamo Stock Farm and for Sparks's convenience provided a stopping point at his headquarters. Sparks then set about making his Truckee Meadows property into a showplace. He imported American buffalo, elk, Persian sheep, and various types of deer. Passengers on the V & T were treated to a parklike atmosphere as they rode through the Alamo property.

The most important animals John Sparks purchased to stock his

Truckee Meadows ranches were purebred Hereford breeding stock. After the turn of the century, Sparks told an interviewer he started in the registered Hereford business in 1875. He probably meant he started experimenting with Hereford bulls for crossbreeding Longhorn cows at that time, but apparently his first attempts to raise his own purebred animals occurred after he developed the Alamo ranch.[23]

In the late nineteenth century, the *Pacific Rural Press* included a gossipy information column for livestock buyers. The January 5, 1889, issue of this publication noted that John Sparks of Reno, Nevada, had 500 steers on feed that winter. The same issue indicated that John Slaver, a Truckee Meadows livestock feeder, had shipped five railroad cars of fat steers to Grayson, Owen and Company in Oakland, California. The steers had been purchased from the ranches of Sparks and Tinnin, and included several thoroughbred, as well as one-half and three-quarters bred Hereford steers.[24]

There are two methods of developing a successful herd of purebred animals. The prospective breeder can purchase a fair number of females of the general type he desires and breed them to a quality foundation sire, select the offspring based on the breeder's standards of quality, and embark on a controlled inbreeding program to fix these qualities in his herd; or, the prospective purebred breeder can go into show-ring sales and spend large amounts of money for flashy, established-name animals, and hope for the best when they are mated. Essentially, the second method involves purchasing from out-breeding populations in contrast to inbred populations. With his ready cash and flair for publicity, Sparks chose the second method. Most of his noted national and international purchases were made in the 1890s after the range operations were reorganized as Sparks and Harrell.

In 1894, Sparks purchased Hereford breeding stock from C.H. Elmendorf of Kearney, Nebraska. This herd made impressive winnings on the stock-show circuit using animals that were the progeny of the foundation bulls, Autocrat and Earl of Shadeland 30th. Previously Sparks had visited Missouri and purchased animals from Fielding Smith and even from the foundation herd of Gudgell and Simpson.[25]

Despite the fact that the livestock industry on the sagebrush/grasslands was still recovering from the disaster of 1889–90 and that the country as a whole and Nevada in particular was in a deep economic depression, John Sparks had money to spend in the early 1890s. He purchased almost the entire first calf heifer crop of the Wyoming Hereford Ranch in 1895 from his old associate of the beef-bonanza days, Alex Swan.[26] These heifers were bred to Luminary 81654, Beau Brilliant 86753, and Patrolman 91594, all quality sires.

In one of their last sales, Gudgell and Simpson and another outstanding Missouri Hereford breeder, James Funkhouser, sold 97 animals at auction for a record average price of $278 per head. John Sparks of Reno, Nevada, was the liberal buyer.[27] Missouri auctioneers must have rubbed their hands with glee when they had John Sparks at the sales ring. Tall, handsome, dressed expensively in western clothes, with a gun under his coat, Sparks himself was an attraction. The national livestock journals, *Breeders Gazette* and *Chicago Drovers Journal* had both carried stories crediting Sparks as "the largest rancher in the west." When the auctioneer turned to Sparks for one more raise on a high-priced bull, the crowd held its breath. How could John refuse, even if he was raising his own bid?

John Sparks ran at least one advertisement in each issue of the *Reno Evening Gazette*, offering his registered Hereford bulls for sale during the late 1890s. Not surprisingly, he enjoyed a very favorable press. In 1898, the editor stressed the need for local Nevada ranchers to buy the surplus Hereford bulls that the Alamo Stock Farm had not been able to sell.[28] Sparks was an innovator in the marketing of registered Herefords. In 1896 he employed the most popular photographer in Reno, E.P. Butler, to take photographs of his prize Herefords. Photographs were not reproduced in newspapers at that time, but the story of a photographic session for bulls was wildly reprinted in Nevada papers.[29]

Sparks loved to take visitors on tours of the Alamo Stock Farm to show his collection of exotic animals and, most importantly, his registered Herefords. The interior of the Sparks mansion was furnished with many curios of his life on the frontier which he enjoyed showing visitors. He once paid $100 for a set of deer antlers locked together in fatal combat. An overzealous servant was assigned the task of polishing the antlers. The servant returned from his assigned task to report to Master Sparks that he had had a terrible time, but he finally succeeded in separating the antlers.[30] Aside from curios and stories, Sparks entertained his visitors with glasses of his special applejack. Editors of the Reno *Evening Gazette* claimed that one drink of Sparks's applejack would make hair grow on a bald head within twenty-four hours.[31] By

1897 the same paper reduced the time necessary for hair to sprout to fifteen minutes.[32]

John Sparks was very good for the developing Hereford industry. He was no midwestern farmer who raised fancy English cattle in a barn; he was the king of the western range, the biggest of the big. A story soon began to circulate in the livestock industry that Sparks started breeding Herefords because he discovered Herefords had survived the winter of 1889–90 much better than Longhorns. Reinhold Sadler, one of the cattleman governors of Nevada, even included this story in his governor's message.[33] The governor reported Sparks was running a 65,000 herd in 1889 and lost 35,000. Of the survivors, 90 percent were Herefords. This kind of publicity was pure gold for Hereford breeders.

In a "compilation of historic facts" about Herefords in America, Sparks was quoted as saying in regard to the winter of 1889–90, "I lost about 30,000 cattle or about 65 percent of my entire herd. The Herefords at the beginning constituted about 40 percent of the whole herd. I found that of the number surviving the second winter, at least 90 percent were Herefords, showing conclusively their superior constitution."[34] It is impossible to calculate, but use of this statement must have resulted in hundreds of thousands of dollars in increased bull sales on the western range. Possibly the most accurate direct quote from Sparks on the losses of 1889–90 was published by *Harper's Weekly.* "We lost that winter, which was a severe one, 35,000 head of cattle and when we rounded up our cattle the following spring, 90 percent of those we found had white faces characteristic of Herefords."[35]

The numbers may still be questioned, but the vital point of this quote is the mention of *white faces* characteristic of Herefords. The animals that survived were probably hybrid offspring of Hereford bulls and Longhorn cows. Survival was a product of hybrid vigor. The Hereford bulls were partial contributors to this vigor as were the genetically diverse Longhorn cows.

The Sparks mansion contained two cases of trophies and medals won by his Hereford show herd. Nevada started a State Fair in 1885. Sparks fitted a show herd which toured the Far West shows such as the California State Fair. There are many references to a bull, Earl of Shadeland 30th, being the champion of the Columbia Exposition of 1893.[36] Earl of Shadeland 30th 30720 was the top herd bull for Sparks's Alamo herd. The bull was exhibited at the Columbia Exposition where it stood third in the sweepstakes class for bulls of any age. At the time, Earl of Shadeland belonged to C.H. Elmendorf. Sparks later bought the bull and gradually referred to the herd bull as "his champion" of the Columbia Exposition. Earl of Shadeland 30th was a tremendous bull who for several seasons was considered undefeatable on the midwestern show circuit.[37]

John Sparks obtained a lot of publicity from Earl of Shadeland 30th, but it was purchased publicity and not a product of his breeding program. Apparently, the best finish in national competition obtained by progeny of his own breeding program was a fourth place finish by a junior yearling bull in the St. Louis World's Fair of 1904.[38]

During the late nineteenth century there was considerable prestige attached to importing Herefords directly from Herefordshire, despite the growth and progress made by American breeders. From 1880 to 1900 some 5,000 head were exported to the United States. This must have left English pastures considerably depleted of stock while enriching English Hereford breeders. Sparks imported eight Herefords, two bulls, and six heifers during this period. One bull was purchased from the Monkton Herd of James Smith located at Pembridge, Hereford. This was one of the first established herds of Herefords in England. The second bull was purchased from the Court House herd of John Price, also located at Pembridge, Hereford. This herd was founded in 1862 and had a brilliant livestock show record in England. The heifers were purchased from A.P. Turner's The Leen Herd, and R. Green's The Whittern Herd as well as the herds of Smith and Price.[39] These eight animals were sufficient to establish his image as an importer. As one exuberant supporter described his herd, "It was capable of competing with the royal herd of Queen Victoria."[40]

Sparks was quoted after the turn of the century that he began importing Herefords in 1893.[41] Apparently his first venture in English cattle occurred in 1897 when he visited Kansas City to see a shipment of Herefords imported from England by Kirkland Armour. He hired George Morgan, a native of Herefordshire, who had helped deliver the Herefords to Armour. Morgan was a quality herdsman who became manager of Sparks's Alamo herd. In 1900 Sparks sent Morgan back to Herefordshire with a bank draft for $10,000 to purchase a quality herd bull.

In 1904, when President Theodore Roosevelt visited the Alamo

Stock Farm, Sparks told him, "Now that you've met my prize bulls, Mr. President, meet my John Bull," and introduced George Morgan. "If you want to ask any questions about the bulls, ask George."[42] This may well have been a tacit admission that review of long pedigrees and reciting English bloodlines was not among the many talents of this cowboy capitalist from Texas. If he lacked technical details in the purebred livestock business, it did not keep Sparks from being active in the political portions of the business. He was an active supporter of the American Hereford Association and an early president of this organization.[43]

John Sparks reached out for technical expertise by hiring William Stevenson whose father had been manager for Gudgell and Simpson. The Stevensons were Scotch cattlemen of the old school. He became manager of the Mayberry Ranch in Sparks's Truckee Meadows holdings.[44] John Sparks then capped his national or international purchases of Herefords by paying the unheard-of price of $10,000 for Dale 66481 in 1900. Dale was among the top sires in the national show circuit.[45]

Always quick to grasp the potential publicity value of his actions, John Sparks protested the values the Washoe County assessor placed on his purebred Herefords in 1901. The assessor valued his bulls at $70 and the Sparks's cows at $50. Sparks raised the assessed value to $500 for bulls and $100 for cows. Sparks appeared before the Board of Equalization and refused to pay until the valuations were raised.[46] As a businessman, John Sparks was no dummy. He succeeded in getting the bulls, which were his marketable product, increased to assessed value more than seven times the original amount while the cows which made up the bulk of his herd on a continuing basis only doubled in assessed value.

In 1889 Sparks became involved in a national scheme to promote Herefords. Kirkland Armour donated, to the promoters of the American Royal Livestock show in Kansas City, a Hereford heifer to be raffled to raise funds for a building program. It was not just any heifer, but Armour Rose 75084, a perfect yearling. She was a product of inbred Anxiety blood, the get of Beau Brummel, Jr., bred by Gudgell and Simpson. Thousands of raffle tickets were sold. A lady held the lucky ticket. Armour bought the heifer back for $1,000 and redonated the animal. The heifer was bought and redonated several times until finally John Sparks and the manager of Marshal Fields's Hereford herd at

Madison, Nebraska, got into a spirited bidding war. When the auctioneer's gavel fell for the last time, Sparks had purchased Armour Rose for the then-record price of $2,500 for a Hereford female. Armour Rose's dam was also related to Beau and as the inbreeding was too close, she was barren and never produced a calf.[47] Although this may appear to be a net loss for Sparks, the national publicity he received added to his bull sales and, undoubtedly, added to what he received for the bulls.

Sparks proved he could still pick a quality steer when he sent a steer named Alamo to the Chicago International. The steer was named grand champion at this show which previously had been the exclusive providence of corn-belt stockfeeders. Armour Packing Company killed the steer that dressed an unbelievable 70.1 percent. Sparks knew when to keep a good thing going. He had Alamo's hide tanned and presented his mounted head to Thomas Marlow, president of the First National Bank of Helena, Montana.[48]

The fortunes of Alamo Stock Farm were on an upswing in 1899. James Wilson, Secretary of Agriculture for President McKinley, visited Sparks's operation.[49] Sparks sent his show string of Herefords to the Kansas City Royal. The Royal hosted the greatest, to date, show of Herefords. Over 500 highly fitted animals were exhibited and 300 head were sold at auction at an average of $317.[50] Sparks returned to Reno to announce his stock were the highest selling animals of the show.[51] By the turn of the century, America had become an exporter of purebred livestock. Sparks became the first Hereford breeder to ship purebred animals to Honolulu, Hawaii.[52]

Sparks's purebred Hereford operation influenced livestock production on the sagebrush/grasslands in three major ways. By supplying quality bulls to ranchers and assembling a critical mass of quality purebred Herefords, Sparks provided a foundation for the Hereford breeder that had an influence long after the Alamo Stock Farm was dispersed; and, Sparks's national reputation enhanced the image of Intermountain agriculture. The last point is obvious from the previously mentioned visits of the Secretary of Agriculture and President Roosevelt to Alamo Stock Farm. The magnitude of this image-building power is apparent when you consider that a half-century before, the Intermountain area was known as a dreadful desert that blocked the way to California or Oregon.

The Alamo herd provided foundation stock for the Whitaker herd of Galt, California, the Jack herd of Salinas, California, Joseph Marsden of Lovelock, Nevada.[53] Many of the top bulls from the Alamo Stock Farm went to Cazer and Son of Wells, Nevada. Prospective Hereford breeders in what had been the most remote backwaters of the United States suddenly had access to bloodlines of the most valuable Herefords in the world.

During the 1890s, the constraints of the environment limited the ranchers' ability to meet the changing pattern in consumer demand for beef. Consumers demanded smaller cuts of tender beef. Midwestern feeders demanded a feeder animal from the western range that was early-maturing and an economical converter of feed to meet. The term "beefs," for huge steers that matured at five years old, disappeared from the livestock industry. Ranchers in the sagebrush/grasslands had few options available to produce younger, more tender beef. Depleted sagebrush ranges cannot produce sufficient forage to fatten two-year-old steers. Only the wetlands or meadows produced forage of quality and quantity to fatten beef. Meadows were in extremely short supply in the Intermountain area. Even if ranchers had the physical environment, irrigated land, and natural meadows on which to finish beef, they needed to change genetic quality toward an earlier-maturing animal.

Just as the "White Heifer that Traveled" helped establish Shorthorns as a breed in eighteenth-century England, livestock shows were the means by which new concepts and trends of breed quality were imparted to nineteenth-century American ranchers. Early maturity became the keynote of the stock shows in both breeding and fat classes. In the late nineteenth century, two calves won the grand championship at the Chicago International over fat steers of all ages. This emphasized the tendency toward smaller and younger beef, and the elimination of age as a barrier to the fitness of a carcass for slaughter.[54]

The thrifty English breeds of cattle provided the genetic material to change the characteristics of the Longhorns. This change worked beautifully, and coupled with the development of refrigeration, turned twentieth-century Americans into confirmed beef eaters, with a thick juicy steak as the symbol of quality. Stopping to analyze the procedure leads to a bizarre sequence of events. How could animals developed under the humid environmental conditions of England, where pastures consisted of a hedge-rimmed acre and animals were kept in barns during winter, become adapted to the arid, fenceless sagebrush/grasslands?

Even after the English breeds were introduced to America, the standards of quality were based on performance in show-rings, and not on performance on the range. Much of the success of the English breeds was due to the linebreeding system. Breeds originated because herdsmen followed the example of Robert Bakewell, fixing characteristics which identified breeds through inbreeding. Within established breeds, certain breeders continued to follow linebreeding to certain bloodlines within the breed. The Anxiety Herefords are a prime example of this technique. Breeders persisted with linebreeding when the advantages were seen in the progeny. Payoff for this continued inbreeding occurred when the English breeds were crossed with the Longhorn cows. The resulting offspring expressed heterosis. This hybrid vigor meant that the animals could survive winters like the 1889–90, be more fertile, and mature earlier. Again, if there had been no Longhorns, the system would never have worked.

One outgrowth of the introduction of English breeds was the equating of quality with uniformity. Obviously, the product was of greater value to feeders if the steers produced from the sagebrush ranges were of uniform size and conformation. But this uniformity was extended to color, horn shape, and other characteristics which contributed little or nothing to carcass quality or feed-conversion efficiency.

Hybrid vigor declined with each generation that improved bulls were used. The first time Herefords were crossed with Longhorns, the response seen in the offspring was tremendous. When the first cross of offspring was bred back to Hereford bulls, the expression of hybrid vigor exhibited by their offspring was proportionately less with each generation. The range herds gradually lost the appearance of Longhorns and assumed the coloring and appearance of the English breeds. All these interacting factors were to have tremendous influence on twentieth-century ranching in the sagebrush/grasslands.

Changes in the levels of technology in the ranch business were tremendous from 1860 to 1900. Longhorns in Texas were free-roaming, free-breeding populations with virtually no interference from man at the beginning of the period. By the turn of the century, ranchers were investing scarce capital in English bulls, worrying about linebreeding and bloodlines, while trying to produce baby beef. At the end of the Civil War, John Sparks was a cowboy driving wild Longhorns up the

213

virgin grasslands of the Great Plains. Ranchers had no capital improvements and the cowboy's tools were restricted to horse, saddle, and rope. By the end of the century John Sparks was hiring imported English herdsmen to give purebred Herefords a bath and curl their hair before parading them in a show-ring!

Notes

1. James MacDonald, *Food from the Far West*, (London: William P. Nimmo, 1878).

2. D.F. Malin, *The Evolution of Breeds*, (Des Moines, Iowa: Wallace Publishing Co., 1923).

3. Ibid.

4. Ralph Bogart, *Improvement of Livestock*, (New York: The MacMillan Co., 1959).

5. Malin, "Evolution of Breeds."

6. Ibid.

7. *Elko Independent*, 14 November 1874.

8. MacDonald, "Food from the Far West."

9. James Sinclair, *History of Shorthorn Cattle*, (London: Vinton and Co., 1907).

10. E.B. Patterson, L.A. Ulph, and Victor Goodwin, *Nevada's Northeast Frontier*, (Sparks, Nevada: Western Printing and Publ., 1964).

11. G.M. Rommel, *American Breeds of Cattle with Remarks on Pedigrees*, (Washington, D.C.: GPO, Bulletin 34, U. S. Bureau of Animal Industry, 1902).

12. Malin, "Evolution of Breeds."

13. Ibid.

14. J.M. Hazelton, *A History of Linebreed Anxiety 4th Herefords of Straight Gudgell and Simpson Breeding*, (Kansas City, Missouri: Association of Breeders of Anxiety 4th Herefords, 1939).

15. Hubert Bancroft, *History of Nevada, Colorado, and Wyoming*, (San Francisco, California: The History Co., 1890).

16. A.H. Sanders, "The Story of Herefords," *Breeder's Gazette*, Chicago, Illinois, 1914. See also, *Chicago Drovers's Journal*, 28 August 1879.

17. Patterson, et al., "Nevada's Northeast Frontier."

18. C.D. Creel, *A History of Nevada Agriculture*, (Reno, Nevada: Max C. Fleischmann College of Agriculture, University of Nevada, 1964).

19. Orville Holderman, "Jewett W. Adams and W.M. McGill: Their Lives and Ranching Empire," M.S. thesis, University of Nevada, Reno, Nevada, 1963.

20. *Elko Independent*, 19 August 1885.

21. Statement prepared by Phillip Earl in support of making the Sparks mansion a historical landmark. Material on file at Nevada Historical Society, Reno, Nevada.

22. *Harper's Weekly*, 20 June 1902, Vol. XLVII:1017–34. Hereafter cited as *Harper's Weekly*, "Nevada."

23. Ibid. Several references such as Patterson's "Nevada Northeast Frontier," give dates in the 1890s for the start of Sparks's registered Hereford venture.

24. *Pacific Rural Press*, 5 January 1889, p. 5.

25. V.S. Truett, "From Longhorns to Herefords: A Pioneer in Nevada's Cattle Industry," and "John Sparks, Cattleman who Became Governor," *Nevada Highways and Parks*, Special Centennial Issue, 1964, pp. 30–31, 50–63.

26. Hazelton, "History of Linebred Anxiety 4th Herefords." See also *Reno Evening Gazette*, 25 November 1895.

27. Sanders, "Story of Herefords."

28. *Reno Evening Gazette*, 4 March 1898.

29. *Reno Evening Gazette*, 31 April 1896.

30. Sam Davis, "The Governor of Nevada: Some Characteristics of the Chief Executive of the Silver State," *Sunset*, IXC, (May 1903):19–20.

31. *Reno Evening Gazette*, 25 November 1895.

32. *Reno Evening Gazette*, 1 May 1897.

33. S.H. Short, "A History of the Nevada Livestock Industry Prior to 1900," M.S. thesis, University of Nevada, Reno, Nevada, 1965.

34. D.R. Ornduff, *The Hereford in America: A Compilation of Historic Facts about the Breed's Background and Bloodlines*, (Kansas City, Missouri: Hereford History Press, n.d.).

35. *Harper's Weekly*, "Nevada."

36. For examples, see Truett, "From Longhorns to Herefords," or Sparks's obituary on file at Nevada Historical Society, Reno. See also Davis, "Governor of Nevada."

37. Malin, "Evolution of Breeds."

38. Sanders, "Story of Herefords."

39. Information on the purchase of Herefords by John Sparks supplied by D. Prothero, The Hereford Herd Book Society, Hereford House, Hereford, England. Sources include the Expert Ledger for 1897 and Auctioneer's Remarks for the sale of the Leen, Courthouse, and Moukron herds.

40. Davis, "Governor of Nevada."

41. *Harper's Weekly*, "Nevada."

42. Truett, "Longhorns to Herefords."

43. "Sparks Obituary," Nevada Historical Society.

44. Truett, "Longhorns to Herefords."

45. *Nevada State Journal*, 28 July 1900.

46. Short, "History of the Nevada Livestock Industry;" and Davis, "Governor of Nevada."

47. Sanders, "Story of Herefords;" and Truett, "Longhorns to Herefords." Armour Rose arrived in Reno on December 18, 1899. *Reno Evening Gazette*, 18 December 1899.

48. Truett, "Longhorns to Herefords."

49. R.I. Fulton, "Camplife on Great Cattle Ranges in Northern Nevada," *Sunset*, Vol. V, (No. 3, July 1900):111–18.

50. Sanders, "Story of Herefords."

51. *Reno Evening Gazette*, 28 October 1899.

52. *Harper's Weekly*, "Nevada."

53. Sanders, "History of Herefords."

54. E.G. Ritzman, "The Development of Livestock Shows and Their Influences on Cattle Breeding and Feeding," *U. S. Bureau Animal Industry Annual Report*, 25:245–56, (Washington, D.C.: GPO, U. S. Dept. Agr., 1908).

IV
THE
LAND
ANSWERS

THE LAND ANSWERS

With the turn of the century came sweeping changes for both land and man in the Great Basin. The cold desert's sagebrush/grasslands were fragile with a limited potential to support life. This inherent potential was often exceeded in the grand experiment to establish ranching. This fragile ecosystem did not bend to accommodate man and his herds, it shattered. Demand for silver depleted the Comstock Lode; just as surely, the demand for rangelands shattered the sagebrush/grasslands.

CHAPTER
12

THE PASSING OF THE OLD GUARD

The 1890s were a strange decade in Nevada. Silver mining was in a terrible slump and a major depression began the decade. The economy and the population of Nevada went down until the question of removing statehood was considered. Further, the winter of 1889–90 had crippled the livestock industry and altered the methods of raising livestock. The bloom was definitely off the sagebrush/grasslands.

After the disastrous winter of 1889–90, the Sparks-Tinnin Company was restructured and incorporated as Sparks-Harrell. John Tinnin left the company to ranch in Nebraska. Jasper Harrell was major owner of the new corporation in partnership with John Sparks. Possibly, Sparks-Tinnin never completed payment for the ranches they had purchased from Jasper Harrell in 1881. The sale had called for mortgage payments of $100,000 annually until the $900,000 purchase price plus interest was paid. Even if payments were made on time, Sparks-Tinnin would still have had payments running on the original mortgage. In a sense, Jasper Harrell was reassuming control of the vast Nevada and Idaho holdings he had founded.

Jasper Harrell continued a vigorous program of water development to bring more land under cultivation on Sparks-Harrell ranches. As time went by, John Sparks apparently drifted away from his ranches in northeastern Nevada and Idaho to concentrate more of his interest in the Alamo Stock Farm at Reno. Jasper's son, Andrew J. Harrell, took an increasingly active role in management of the Sparks-Harrell ranches during the decade of the 1890s.

The depth of the economic situation in Nevada is illustrated by a resolution adopted by the Sparks-Harrell Company at their annual meeting held in Visalia, California, February 7, 1895: "That all persons employed by the superintendent, any foreman, or any employee having authority to engage the services of any person in behalf of this Company, on or after the 15th day of February 1895; that the salary of said new employee, or person hired, shall not exceed the sum of thirty ($30.00) dollars per month."[1]

Despite his interest in the Alamo Stock Farm, John Sparks was still president of Sparks-Harrell Company midway through the decade of the 1890s. At that time, and in subsequent accounts, John Sparks was considered to be the driving and managing force in the Sparks-Harrell Company.[2] A letter written by A.J. Harrell in 1896 to his cousin, Louis Harrell, shed some light on the operations. "I am disappointed that Harris and Duncan did not gather more than 1,100 head of cattle, as I thought there were a great many more than that number there; when you get this, write me how many cattle you think are left on that range, and if the boys said they got a good clean gathering. And you go to the desert as we talked it over, and keep a good watch on the cattle we have left there, and organize an outfit big enough to make a clean gathering next spring; do all this without any further orders, for we may not have a chance to instruct you further before work begins. I have not heard from Sparks since I saw you, but think you will see him by the time you get this, and when you do, talk over with him what I have instructed you to do, and if he disapproves of it, do as he says; but if you do not see him, carry out my instructions until you get further orders."

This letter suggests that John Sparks was still a power in Sparks-Harrell, but that he was not in close enough touch with ranch operations to plan and direct such important events as roundups.[3]

Some basic constraints began to pressure Sparks-Harrell. The winter of 1889–90 made it obvious that hay reserves were necessary to guard against excessive winter losses. Raising hay is closely tied to irrigation. On balance, it would seem that the effect of the "white winter" would strengthen the influence of landowners who controlled irrigation water. There was, however, a livestock production pattern which avoided the necessity for hay production. This method dictated the use of sheep. Sheep are more efficient browsers, require less water per equivalent animal unit, and can utilize winter range while depending on snow for water. Also, less capital was required for sheep than cattle. If hay production requirements for cattle were included, it was ten times as costly to go into the cattle business as the sheep business. Cull ewes had less residual value than cull brood cows, but their initial cost was proportionately less. Many range sheep operators owned no land and were completely nomadic.[4]

Cattle ranchers faced competition not only on winter ranges, but also on higher elevation summer ranges. Summer range became the most limiting factor in livestock production. Not only were cattle and sheep degrading the same overgrazed range, they were destroying the vegetative cover of the watersheds that provided the runoff which supported agriculture and hay production.

Sparks-Harrell faced continued pressure from the Northeast. The spread of range sheep appeared to Sparks-Harrell to be a threat to "their" rangelands. More than 90 percent of their rangeland was public land, and major partners were absentee landlords for the land they did own. The mountain corridor across southeastern Idaho which provided a route for the California Trail also provided the environment for the spread of relatively small sheep operations. Many of those running sheep were members of tightly knit Mormon communities.

In perspective, Jasper Harrell had virtually founded ranching in northeastern Nevada and south-central Idaho. If he could have obtained title to all the land where he ran livestock, he doubtless would have. But his and his partner's possessory claim to the range now lacked legal status. Sparks-Harrell's principal hold on the rangelands was through force and prior arrival.

This threat of competition forced Sparks-Harrell to draw a line across the Goose Creek Basin, with sheep not to be permitted west or south of this "deadline." By mid-1895, sheepmen became more aggressive, and took flocks across the dividing line, inviting consequences. Sparks-Harrell then hired "outside men" for their ranches. Among the outside men hired as gunmen whose jobs involved physical violence were Jack Davis, Billy Majors, Fred Gleason, and William Majors.[5] Instructions through James E. Bower, superintendent, and Joe Langford, range foreman, were "Keep the sheep back. Don't kill, but shoot to wound, if necessary. Use what measures you think are best. If you do have to kill, the company will stand behind you. There is plenty of money and backing, and the company won't desert you regardless of what happens." Today this action seems unbelievable, but it should be viewed in context of the 1890s. Captains of industry resorted to similar tactics in fighting the labor movement.

John Sparks had had previous experience with the Wyoming Stock Grower's Association, which employed detectives to help cope with rustling. This eventually led to the Johnson County Range War. In Nevada Sparks employed T.M. Overfelt, previously employed by the Wyoming Stock Grower's Association as a range detective. Overfelt was killed near Elko by a runaway team under rather suspicious circumstances. As a Texas Ranger, John Sparks had fought Comanche Indians, defending early settlers' homes and farms. In 1875, on Wyoming's North Platte River, he had established the frontier ranch of that section of the county and defended it against the Sioux. That same year, when an outbreak of the Cheyenne and Sioux occurred, Sparks was given by territorial Wyoming Governor Thayer a captain's commission with guns and ammunition to protect the settlers.[6]

A central character in the drama that unfolded on Deadline Ridge was Jack Davis. He was an unlikely looking gunman. Short, slight of build with ruddy face and a sandy mustache, Jack Davis was likable, well-mannered, and kind. He was a complex personality who compulsively talked himself into trouble. He endlessly bragged about his prowess as a gunman, which he called "cutting it in smoke." Despite his endless stories, little was known of his background. If he could be believed, he had been born in four different states, fought Apaches in Arizona, revolutions in South America, and worked with Cecil Rhodes in South Africa. Also, he had been a miner in the early 1890s working in the Silver City district of Owyhee County, Idaho. Rumors of a diamond strike in some nearby hills prompted Davis to go and seek his fortune as a prospector. The diamond fields proved an illusion, but Jack seldom tired of telling what he would have done if he had found the diamonds. A cowboy friend christened him "Diamondfield," and the name stuck.

When Diamondfield Jack started patrolling range for Sparks-Harrell, he spared no words telling isolated herders how dangerous he was, and what he would do if they crossed the deadline. His pattern was to ask the herders who they worked for. After they answered, he told them if they worked for a different outfit he would have had to shoot them. The sheepmen soon discovered that Jack was always gunning for a sheepman who was somewhere else at the moment. Often in the process of threatening some poor herder, Diamondfield Jack might break off the argument and have supper with his adversary if the grub smelled good.

A prominent Idaho sheepman, Bill Tolman, boldly rode up to a Shoshone Basin Sparks-Harrell lineshack and confronted Jack with a rifle. Diamondfield engaged Tolman in a long argument, after which he drew his .45 and shot Tolman through the shoulder. Diamondfield Jack then treated the fallen sheepman's wound, and made him as comfortable as possible. He later transported Tolman to a sheep camp so he could be moved to Oakley, Idaho, for treatment. Sight of the wounded sheepman being carried home was enough to discourage other herders from crossing the deadline.

Apparently Diamondfield Jack realized he was in trouble, as he collected his pay and headed for Wells, Nevada. In the warmth of Fisher's saloon and Alice Wood's palace of pleasure he was soon happily telling everyone he had been "up in Idaho shooting sheepherders." On February 16, 1896, a sheepherder on Deep Creek noticed scattered bands of sheep. When he went to investigate, he encountered a grisly scene; two young Mormon sheepherders, John Wilson and Daniel Cummings were found shot to death in their wagon. Diamondfield Jack had returned to work for Sparks-Harrell. Not much imagination was required on the part of Cassia County sheepmen to make Diamondfield Jack the prime suspect.

Almost a century after the killings, two sensational trials, pardon board hearings, newspaper and historical articles and books, it still is a volatile issue in the Intermountain area as to who killed Wilson and Cummings. Based on their own confessions and evidence, the killings were done in self-defense by James E. Bower and Jeff Gray. Bower was the general manager of Sparks-Harrell. He was the cowboy who, upon first glimpse of the Snake River Valley grasslands, had hurried back to Jasper Harrell with news that helped found the Harrell ranching empire. Bower had helped build the first schoolhouse in the area and organized the first Sunday School classes. He was also an independent cattleman in Idaho. Jeff Gray was a local boy who had grown up with the area sheepherders.

From sworn affidavits, it was learned that Bower and Gray stopped at a sheep camp on the morning of February 4, 1896, to question the two herders about their intentions of moving onto Sparks-Harrell range. As all four sat around the stove in the sheep wagon, hot words were exchanged, and the men grappled. When it was over, the herders were shot. Thinking only one man was wounded, and that neighboring sheep camps would have heard the shots and be in hot pursuit, Bower and Gray rode away. Bower and Gray told several cattlemen about the incident before and after they learned that both men had been killed. Gray confided to A.D. Norton, the pioneer livestock man of Rock Creek, Idaho. Bower presumably told John Sparks and Jasper Harrell.

Confessions to the crime, in affidavits, did not come forth until October 13, 1898, after Diamondfield Jack had twice been tried and sentenced to hang! Both times he waited under the gallows before a rider galloped in with reprieve from the governor of Idaho. The trials featured top defense lawyers paid for by Sparks-Harrell, and special prosecutors rumored to be sponsored by the Mormon Church. Trials and resulting publicity helped establish the political career of William E. Borah, the "Lion of Idaho," as a United States senator.

Diamondfield Jack based his defense on the timing of his physical presence at various ranches on Salmon Falls Creek. He was seen at the Brown Ranch in Idaho and at the Sparks-Harrell Boar's Nest Ranch on the Nevada side of the stateline by many witnesses. But the question became, "Did Diamondfield Jack ride at a normal speed from ranch to ranch or did he dash to Deep Creek, kill the sheepmen and continue to the Boar's Nest Ranch?"[8]

Diamondfield Jack's trials were held in Cassia County, Idaho, with jury composed of sheepmen. Conviction was expected. Following the second trial was a long period of appeals, stays of execution, and meetings at the Idaho State Pardons Board. After the Bower-Gray confessions were made public, it was obvious Diamondfield Jack was being held for the wrong crime. On December 17, 1902, the Idaho Board of Pardons granted Diamondfield Jack a full pardon. He had spent six years in prison for a crime he apparently did not commit. He had very nearly talked himself to death.

Diamondfield Jack's case provided many rumors. One was that the Mormon Church helped finance his prosecution. After Jack's pardon, there were persistent rumors that the pardon cost John Sparks a small fortune. Sparks had made it no secret that the company financed the cost of Diamondfield's defense and appeals. If bribery of the pardon board was involved, it has escaped discovery by historians despite persistent investigations. Apparently Diamondfield Jack was paid well by Sparks-Harrell for his six years in prison. After being released from the Idaho State Prison in Boise, he reappeared in south-central Nevada during the Goldfield and Tonopah booms in the early twentieth century, and made money in mining and land speculation. He also proved useful to the Wingfield-Sparks interest in fighting labor unrest in the mines and mills. He developed a mining camp of Diamondfield, Nevada. When the boom wore off in central Nevada mining, he became a successful real estate developer in southern California. He died after he was hit by a taxi while on a gambling trip to Las Vegas, Nevada, in 1949.

For the livestock industry, the Diamond Jack case was symptomatic of the problems the industry was having. The cattle culture had been operating in excess of potential of the resources. Basic forage resources had been severely depleted. Coupled with the severe winter of 1889, this had nearly destroyed the industry and forced redesign of forage utilization, centered on the hay production.

Established ranchers were caught in a paradox. If they owned the land, the taxes were greater than the potential income under state and national economic conditions.[9] If economically you could not afford the necessary rangeland to support operations, the only alternative was to protect your possessory right on the public rangeland. Sparks-Harrell were protecting these rights when they hired outside men such as Jack Davis.

The same desire expressed through political action led pressure from the large ranches to establish National Forest on the higher mountain ranges in the early twentieth century. Grazing permits were allotted on the basis of history of use, and ownership of commensurate property to support the livestock when not on the National Forest. Both of these requirements favored established ranchers. Essentially, the federal government was eventually to provide a de facto recognition of the possessory interest of the ranchers. The National Forests were at least a decade away in the 1890s.

Expansion of the range sheep industry offered an alternative that required much less capital, and negated the land/water control aspect. Rise of the range sheep industry was possible because the cattle ranchers lacked legal avenues and capital to acquire title to the extensive acreage essential to cattle operations in the sagebrush/grasslands. There was one obvious solution for Sparks-Harrell to meet the competition from range sheep. That solution was to go into the sheep business. This method was employed by the Utah Construction Company for the first half of the twentieth century on the former Sparks-Harrell ranches.[10] Why did Sparks-Harrell fail to go into the sheep business? Lack of knowledge about the sheep business may have contributed, but traditional prejudice against sheep and sheepherders was probably a bigger factor.

The depth of the ill feelings that existed between established cattle ranchers and sheep operators is apparent in the Diamondfield Jack case and from other sources. For example, Governor Sadler in his address to the state in 1896, stated, "With the exception of a comparatively few million acres, the entire State of Nevada must remain a stock raising state and the greatest benefit which she could not [sic] receive at the hands of Congress would be some manner of legislation looking to the preservation of her ranges to prevent them from being laid waste by foreigners—men who are not citizens, never intend to become such, and who use our state only to ruin it, and filch from our people their national heritage."[11]

Foreigners were very much a part of the controversy. There were several large cattle operations which branched into the sheep business with scarcely a comment from their neighbors. It was the new sheep outfits, often owned by Basque herders, that especially raised the anger of cattle ranchers. Many of these new sheep outfits were completely migratory without base property in Nevada. Even government publications such as Griffith's survey of the northern Great Basin at the end of the 1890s slurred the religion and national origin of the migratory sheep outfits.[12]

John Sparks was always ready to enhance his reputation and standing. He used the wildlife resources of the Sparks-Harrell rangelands to entertain and curry favors with a variety of powerful bankers, judges, and financiers.[13] In late August guest hunters from Cheyenne, Salt Lake, Los Angeles, and San Francisco dropped off the train in Wells, Nevada, to be ushered to the Sparks-Harrell ranches. The hunters traveled in style for Sparks-Harrell provided a cook, camp tenders, a horse wrangler, and forty extra riding horses.[14] The big-game hunters were after mule deer. Bucks had antlers still in the velvet in late August, and it is difficult to see how the hunters kept meat from spoiling in the hot weather typical of early autumn. There were abundant sage grouse to shoot around the stringer meadows, and the more venturesome hunters tried for pronghorns in the valleys.

Always the extrovert, John Sparks packed his special rifle "Alcade" and displayed notches in the stock which represented deer, bear, elk, and buffalo he had shot. Alcade was a Sharp 50-caliber rifle or "buffalo gun." Once he demonstrated how well he could shoot the old rifle by knocking down a buck at a reputed 600 yards.[15] One gets the impression that Sparks enjoyed playing the role of king of the sagebrush ranchers before his audience of powerful and influential friends. He would often serve a special Texas breakfast consisting of a baked bull's head, split so the guests could scoop out the brains.

The hunters on Sparks's trips had good luck with sagegrouse, often killing far more than they could consume. The sagegrouse is native to the sagebrush/grasslands, and apparently there were enough meadow environments left in the sagebrush/grasslands in the 1890s to support sagegrouse chicks, despite the general abuse of the rangelands. Mule deer were not abundant, even in the high mountains of extreme north-eastern Nevada. On some years, the entire party was lucky to get a single shot at a buck. Shrubs were increasing on the sagebrush/-grasslands, especially in the aftermath of the winter of 1889-90. Mule deer populations were responding to this increase in their environment potential, but populations had not yet increased to the point where deer were abundant. On one trip Sparks blamed the lack of deer on the presence of Indians on his range.[16]

Intermountain area ranges were influenced during the decade of the 1890s by the recurring story of drought in California and the movement of cattle across the Sierra Nevada. California experienced varying degrees of drought from 1895 to 1900. In June of 1895 starving cattle were arriving in Reno from California, and by midsummer John Sparks was reporting that forage was extremely short on Nevada rangelands.[17] California ranges were again dry in 1896. The problems of California stockmen were complicated by the outbreak of Texas fever in southern California. The Secretary of Agriculture, J. Sterling Morton, clamped a federal quarantine on the shipment of cattle out of Nevada. No one knows what the motives of Nevada stockmen were, but they hailed this quarantine as a great and good thing. Apparently everyone was interested in protecting the health of his stock, but the quarantine also happened to keep the giant firm of Miller and Lux from shipping thousands of cattle from California to Nevada where they traditionally summered.[18] The federal government lifted the quarantine in late winter and Nevada cattlemen immediately put pressure on Governor Reinhold Sadler to declare a state quarantine against California cattle. The Governor was a cattleman from Eureka County, Nevada, and responded favorably to the request. Governor Sadler immediately found himself in hot water. J.H. Budd, the governor of California, personally applied to Sadler to lift the quarantine to save starving cattle.

Miller and Lux was a powerful firm in California with great political influence. Virtually every cattleman in Nevada signed petitions to keep the quarantine in effect. Surprisingly, John Sparks was quiet about the controversy. John Sparks knew all about Texas fever, as his herd had carried it from Texas to Virginia in the 1860s. As the quarantine dragged on to the turn of the century, Sparks received critical publicity for importing directly from Texas purebred, but uninspected, Jersey cows to Alamo Stock Farm.

The quarantine controversy heated up in 1899 when Miller and Lux requested permission to ship 10,000 head of cattle through western Nevada. The cattle were to stock a ranch in southeastern Oregon, which the company had purchased from the Sharon Estate. Miller and Lux reported that the cattle had been gathered off the California winter range in such poor condition that they did not have the strength to be driven to their new range. The company wished to ship the cattle to Reno on the Southern Pacific and reload them in narrow-gauge cars for shipment north on the Nevada California and Oregon Railroad. Nevada papers immediately jumped on Governor Sadler, even before he replied to Miller and Lux. Essentially, the quarantine became a method for the established ranchers who lived in Nevada to extend a greater degree of de facto control over the public rangelands. The ranchers even attempted to extend the quarantine to animals that were not susceptible to Texas fever, especially California sheep. In 1899, the Governor of Idaho wrote Governor Sadler requesting information on the quarantine as he proposed to extend it to Idaho as suggested by that prominent cattleman, John Sparks, who happened to be visiting his office.[19]

John Sparks continued as a national figure and enlarged his reputation as a spokesman for the livestock industry.[20] He served on the public lands committee which tried to lobby for a federal policy of the leasing of public lands by ranchers. Sparks sponsored a delegate, Dr. J.E. Stubbs, president of the University of Nevada, whose attendance was considered quite unusual at the time. The Agricultural Experiment Station of the University of Nevada was founded in 1887 under the provisions of the Hatch Act.[21] The concept of universities conducting agricultural research was conceived during this period.

Among the delegates at the 1898 National Stock Grower's Convention was the familiar name of John Tinnin, listed as a delegate from South Dakota, but residing at Gordon, Nebraska. The Nevada delegates read as a who's who of the nineteenth-century livestock industry: Col. Hardesty, J.R. Bradley, N.H.A. Mason, J.J. Altube, and A.C. Cleveland. By 1900 there was a change apparent in the delegate structure of the National Stock Grower's Association. Members of the State Wool Grower's Association were elected delegates to the national meeting of the stock growers. W.H. Poulton, Oakley, Cassia County, Idaho, was a delegate from the state Sheep and Wool Growers.[22] This association may have helped heal some of the wounds created by the Diamondfield Jack affair.

Among the presentation made by delegates at the 1898 meeting of the National Stock Grower's Association was one by William Byers, an obscure rancher who spoke from the heart and touched the soul of the range livestock industry that had experienced hard winters on the plains and the winter of 1889–90 on the sagebrush/grasslands. Speaking under the title "The Humane Treatment of the Range Stock," Mr. Byers stressed the value of conserving forage for use in wintering cattle. He pointed out that one-half ton of hay made a yearling or two-year old worth five to eight dollars more in the spring than if it was wintered on the range without supplemental feeding. He suggested the cost of hay production on the ranch could not be over two to three dollars per ton. Byers concluded his talk, "*Whenever animals are under man's control, it is his duty to see that they do not suffer from any cause which he is able to remove.*"[23]

The operational style of Sparks-Harrell began to change during the mid-1890s. After 1896 the Sparks-Harrell roundup wagon no longer operated on the Owyhee Desert. This vast area contained much winter range and had been a stronghold of Jasper Harrell's operations in the 1870s.[24] Competition from smaller ranchers in Idaho, especially the Bruneau Valley, who had expanded after the winter of 1889–90, forced Sparks-Harrell to reduce the extent of their operation. While John Sparks was increasing his political activity and interest in the Alamo Stock Farm, Andrew J. Harrell was enlarging the irrigation ditch network on the company land and bringing additional land into hay production.

A.J. Harrell was described in some accounts of the Sparks-Harrell operation as a Bay Area butcher and a California playboy who John Sparks tolerated as he guided the fortune of the company.[25] In actuality, A.J. Harrell, with a business college education and experience gained in ranching, real estate, and banking, was the driving force behind Sparks-Harrell during the 1890s. Behind A.J. was the flint-hard figure of Jasper Harrell. In 1891, Jasper and A.J. sold the Harrell & Son Bank of Visalia to the Producers Bank, and A.J. Harrell then devoted full time to the firm's Nevada interest.[26] In 1899 A.J. Harrell moved his wife and two children from Visalia to Palo Alto so he could be closer to San Francisco, the financial and cultural center of California.

Harrell's influence on the Sparks-Harrell operation was apparent when California banker, Jackson Graves, visited the ranches in the mid-1890s. The H-D Ranch had 1,800 acres of irrigated meadow and alfalfa. The Hubbard and Vineyard ranches both had substantial pole buildings with dirt roofs. At the Hubbard, 3,000 acres of sagebrush had been converted to irrigated fields. At all the ranches, immense stacks of hay were seen. Some of the stacks had been carried over from previous seasons as insurance against another winter like 1889–90.[27]

The year 1901 marked the end of the introduction and expansion period for cattle in the sagebrush/grasslands. John Sparks sold his interest in the Sparks-Harrell ranches to A.J. Harrell. The April 4, 1901, issue of the *Reno Evening Gazette* headlined "The Sparks Sale—Over a Million Paid for his Cattle and Ranch Property." The paper indicated that the deal had been brought to a conclusion in Salt Lake City five days previous to the announcement. John Sparks had sold to A.J. Harrell of Visalia, California, 20,000 head of range cattle along with a one-half interest in 700,000 acres of land and a lease on 700,000 additional acres. Sparks refused to discuss the purchase price with reporters, but it was rumored that he received $500,000 in cash and 12,000 acres of Texas cotton land for his half-interest in Sparks-Harrell. Sparks was quoted by Salt Lake papers, "I plan to devote some attention to raising cotton on my little 12,000 acre Texas ranch. . . . Cotton is King you know, and if I raise enough of it, I may make some money—there is no telling."[28] The firm continued until 1908 as Sparks-Harrell.[29] In his widely quoted interview published in *Harper's Weekly* in 1902, Sparks strongly implied he was still owner of the Sparks-Harrell ranches.[30] To suddenly become publicity-shy seems very out of character for John Sparks. It had been ten years since he had refinanced Sparks-Tinnin into Sparks-Harrell. Although he told Fulton he yielded to a tempting offer in giving up Sparks-Harrell, it may not have been by choice. Considering his later financial problems, John Sparks probably was unable to pay his mortgage debts to the Harrells and lost his ranching empire.[31]

On May 13, 1901, Jasper Harrell died. A.J. Harrell inherited sole ownership of 175,000 acres in the central Great Basin, and 30,000 cattle, which ranged on three million acres.[32] In the 1880s, this ranching empire had, by some reports, 150,000 head of cattle. This is one of the most striking statistics of how overstocked the ranges were in the 1800s. In 1901, brood cows were getting one-half their forage requirement from irrigated lands in the form of hay and crop aftermath. In the 1880s, a reported five times as many cows were dependent on the range.

While President Stubbs was busy selling university research at the National Stock Grower's Association, there were other moves underway to enhance the western rangelands. Secretary of Agriculture Wilson announced in 1898 that the U.S. Department of Agriculture (USDA) had sent Professor Niels Hansen from Brookings, South Dakota, to central Asia to collect plants for revegetating semiarid areas in the West.[33]

Plant explorations were also being carried out closer to home. The USDA sponsored a survey of the western range to try and determine the nature of the resource. David Griffith, G. Vadey, F. Lamson-Schriber, A. Nelson, and G. Smith for the USDA, F.H. Hillmand and P.B. Kennedy of the University of Nevada, and H.T. French of the University of Idaho began to describe the nature and extent of the grazing resource.

In 1904, A.J. Harrell was experimenting with grasses from the Russian steppe on his properties in Nevada.[34] This was an attempt to fill the niche left open by the destruction of the native perennial grasses. The grass was described as a "recent import from Russia." Although it is unlikely at that early date, the grass may have been crested wheatgrass introduced by Professor Hansen. By the mid-twentieth century, millions of acres of severely degraded sagebrush rangelands were seeded to this species.

Native shrubs, especially sagebrush, had partly preempted the environmental potential released by the destruction of the perennial grasses. The sagebrush communities became stark, shrub-dominated landscapes without sufficient understories to support anything but marginal livestock production. Most importantly, the sagebrush-dominated communities were extremely stable for the life of the shrubs.

There were not enough herbaceous understory species to carry fire through the shrub communities. The shrubs did not have the ecologic amplitude to fill the potential of the communities. Biological near-vacuums do not last forever, and the longer the vacuums last, the greater the dynamics when filled.

One development was already in process. In the Red Desert of Wyoming a new plant species had been discovered. It was a spiny, coarse herb that uprooted when it was mature and tumbled across the landscape spreading seeds. Known as Russian thistle, it had been accidentally introduced to South Dakota in the 1870s and was to become the first of the alien annual weeds to spread across the sagebrush/grasslands.[35]

With the death of Jasper Harrell soon after the turn of the century, the days of the empire he founded were numbered. The remaining principal characters did not last the decade. John Sparks had a spectacular rise in Nevada politics and was twice elected governor after an unsuccessful race for the Senate. Sparks, the colorful promoter, was caught in a mining promotion and lost a great deal of money for himself and for his friends.[36] The last acts in his action-filled life occurred during the continued strike and labor controversy between Tonopah and Gold-field mine owners and the miners' union. He got federal troops from President Roosevelt to control the area. He arose from his sickbed in midwinter and rode sixty-five miles to do what he could on the scene, but was rebuffed by Teddy Roosevelt who called back the troops. Broke and broken, Honest John Sparks went home to the Alamo Stock Farm and died of Bright's disease; and, as some said, a broken heart. The year was 1908 and it also saw the passing of A.J. Harrell.

Banker Jackson Graves experienced the passing of both friends. He was visiting an estate at Lake Tahoe, where it was becoming quite fashionable to have a summer home. He went down to the dock to see the steamer, *Talac*, the pride of the Lake. As the boat pulled away in the mist he saw a forlorn figure, wrapped in a heavy coat standing on the deck whom he recognized as A.J. Harrell. When Graves returned to Los Angeles, he was doubly shocked to hear that Harrell had passed away. Later that year Jackson Graves visited Reno and drove south of town to pay his respects to John Sparks's widow. He found the front door of the beautiful Sparks home boarded shut and the registered Herefords and exotic animals were gone. John Sparks had died virtually bankrupt.

The Sparks-Harrell ranches passed through intermediary ownership before being sold as a block to the Utah Construction Company under whose ownership it remained until after World War II.

Writing a decade after Sparks's death, John Clay listed six reasons for John Sparks's financial failure: (1) low stock prices; (2) winter losses; (3) distance from market; (4) politics; (5) mining; and (6) purebred cattle.[37]

John Sparks had been the dominant figure in livestock in the Great Basin during the last quarter of the nineteenth century. Two years after his death, his beautiful home at Alamo Stock Farm was sold to William Moffat, the dominant figure in the livestock industry for the first half of the twentieth century.

The pioneers of the livestock business passed away one by one. A few characters in the drama, such as Indian Mike and Henry Harris, lapped over into the twentieth century.

The pristine vegetation of the sagebrush/grasslands was gone. It was still possible to return to the pristine plant communities because the spread of alien weeds had not yet occurred. The range was in trouble, but no one could correct the problem.

There are now vast areas of sagebrush/grasslands that are in relatively good condition that could greatly benefit from grazing management. At the end of the nineteenth century, there were areas at least as abundant as those of today. What if the federal government had sold or leased the sagebrush rangelands to established ranchers? Probably results would have been mixed, but there are interesting results from two sources. One is a huge block of land on Golliher Mountain that Sparks-Harrell obtained from state select land. The other is known as St. John's Field, north of Battle Mountain, Nevada.[38] Both areas have been in private ownership since the nineteenth century, and consist of large blocks of upland range. They are both examples of near-pristine sagebrush/grasslands. This does not mean to imply that private ownership would have resulted in better range conditions across the landscape, but it is food for thought.

Expansion of the cattle industry into the sagebrush/grasslands was a grand experiment, where herdsmen boldly ventured into an environment, then considered beyond the potential of agricultural enterprise. One result was the birth of a system where cattle graze on extensive rangelands, except for fall and winter, when they eat hay or graze crop aftermath. Hay is essential for this system, but it is produced on a fraction of the landscape that can be irrigated. The second result of this experiment was that the sagebrush/grasslands vegetation born in the wild climatic fluctuations of the Pleistocene, and scantily nurtured by the post-Ice Age aridity of the Intermountain area, was destroyed in forty years of domestic livestock grazing.

Two basic factors contribute to destruction of the sagebrush/grasslands. First, the landscape-dominating shrub, known as big sagebrush, was protected from excessive grazing by the essential oil content of its herbage. Second, the dominant perennial grasses in the forage base were reproduced largely from seeds. If excessive numbers of livestock utilized the sagebrush/grasslands, grass disappeared and shrubs increased. If the grasses were not given a chance to redevelop their carbohydrate reserves to permit flowering and seed production, their chances were slim. Under pristine conditions, no concentrations of large herbivores grazed the native grasses. The regeneration system of the native grasses was adapted to occasional stand renewal under conditions of exceptional precipitation.

The nineteenth-century method of determining stocking rates for the new rangelands of the West was to increase stock until death losses became unacceptable. Under this approach, native grasses needed to establish new seedlings, declined annually regardless of the growth potential of the year. Native herbaceous vegetation failed to adapt to the changing stand renewal system. The speed of evolutionary change did not equal the rate of environmental degradation. Native shrubs responded favorably to the destruction of herbaceous vegetation, but exploitation levels of the two life-forms were sufficiently diverse that the shrubs could never completely preempt the released potential. This environment was open to the introduction of plant material with genetic potential in excess of the native perennial grasses. The annual destructive grazing of the herbaceous vegetation put a premium on annual growth forms, especially annuals that had breeding systems that responded rapidly to changing environments to permit the evolution of adapted progenies.

The necessity to produce and harvest hay had profound sociological influences. Cattle ranching started as a very extensive type of agriculture with minimum labor. Indians of the Intermountain area became a major source of this labor, both as year-round cowboys and seasonal hay hands.

The turn of the century was a time of change on the sagebrush/grasslands. The National Forest and federally funded reclamation projects were to come into being in the next decade. Alien weeds such as cheatgrass had not yet reached the degraded sagebrush ranges. The old guard of ranchers, like Sparks, Harrell, Cleveland, Bradley, and Hardesty, were passing away. The cowboys who came astride horses and tried to avoid labor on foot were fading. The cowboy who survived became a ditch mucker and hay hand aside from his work with cattle and horses. The ranchhand-cowboy who came to town to drink and carouse was out of step with changing standards of social behavior.

One singular thing about the open range era of livestock was that it lasted only briefly in its pure form. But the image of the cowboy as a free-living spirit of the plains became fixed and long-lasting.

Notes

1. Letter to Louis Harrell dated February 12, 1895, and signed by John Sparks, president, and A.J. Harrell, secretary, Sparks-Harrell Company, Visalia, California. Copy of letter furnished by Newton T. Harrell, Twin Falls, Idaho.

2. E.B. Patterson, L.A. Ulph, and Victor Goodwin, *Nevada's Northeast Frontier*, (Sparks, Nevada: Western Printing and Publishing, 1969).

3. Letter to Louis Harrell dated December 28, 1896, and signed by A.J. Harrell, Visalia, California. Copy of latter furnished by Newton T. Harrell, Twin Falls, Idaho. Harris and Duncan were Henry Harris and Tap Duncan, the Sparks-Harrell wagon or roundup foremen. The desert is probably the Owyhee Desert of south-central Idaho.

4. B.H. Sawyer, *Nevada Nomads*, (San Jose, California: Harlan-Young Press, 1971).

5. Most of the data on Diamondfield Jack is based on the book, *Diamondfield Jack*, by D.H. Grover, (Reno, Nevada: University of Nevada Press, 1968). Exceptions are materials collected from sources Grover did not use; they are cited as separate sources.

6. V.S. Truett, *On the Hoof and Horn in Nevada*, (Los Angeles, California: Gehrett-Truett-Hall, 1950). Information on Sparks's military career in Wyoming from H.H. Bancroft Collections, Bancroft Library, University of California, Berkeley, California.

7. Patterson, et al., "Nevada's Northeast Frontier."

8. Louis Harrell told his sons he saw Diamondfield Jack arrive at the Boar's Nest Ranch and his horse did not appear to have been ridden hard. This was brought out in court by other witnesses.

9. E.O. Wooton, *The Public Domain of Nevada and Factors Affecting its Use*, (Washington, D.C.: GPO, Tech. Bull. No. 301, U. S. Dept. of Agric., 1932). Wooton recognized the relationship among taxes, production, and land ownership and the disproportionate tax load that agricultural landowners carried during the late nineteenth and early twentieth centuries.

10. To keep competition away from their cattle ranges, the Utah Construction Company fully occupied available winter range with over 40,000 ewes. See N.J. Bowman, *Only the Mountains Remain*, (Caldwell, Idaho: The Caxton Printers, 1958).

11. Papers of Governor Sadler are on file at the Nevada Historical Society, Reno, Nevada.

12. David Griffiths, *Forage Conditions on the Northern Border of the Great Basin*, (Washington, D.C.: GPO, Bureau of Plant Industry, Bull. No. 15, U. S. Dept. of Agric., 1902).

13. R.I. Fulton, "Camplife on Great Cattle Ranges in Northern Nevada," *Sunset*, Vol. V (No. 3, July 1900):111–18.

14. *Reno Evening Gazette*, 21 August 1899, indicated that Sparks boarded the train in Reno for his annual hunting trip.

15. Fulton, "Camplife." B. Abbott Sparks, Jr., located Alcade in California in 1984.

16. J.A. Graves, *California Memories*, (Los Angeles, California: Times-Mirror Press, 1930).

17. *Reno Evening Gazette*, 3 June 1895; also, 3 July 1895.

18. Papers of Governor Sadler are on file at the Nevada Historical Society, Reno, Nevada.

19. Ibid.

20. C.F. Martin, *Proceedings of the National Stock Grower's Association Convention*, (Denver, Colorado: Smith-Brock Printing, 1898).

21. C.D. Creel, *A History of Nevada Agriculture*, (Reno, Nevada: Max C. Fleischmann College of Agriculture, University of Nevada, 1964).

22. W.H. Poulton was the granduncle of C.E. Poulton, noted range scientist, educator, plant ecologist, and specialist in remote sensing.

23. Martin, "Proc. National Stock Grower's Association."

24. Adelaide Hawes, *The Valley of Tall Grass*, (Bruneau, Idaho: Private Printing, 1950). Remarks about the end of Sparks-Harrell operation on Owyhee Desert from a 1937 letter that Mrs. Hawes's father sent to the Regional Forester, Forest Service, U. S. Department of Agriculture, Ogden, Utah. Apparently Louis Harrell never organized the last big roundup as instructed by A.J. Harrell.

25. See Patterson, et al., "Nevada's Northeast Frontier."

26. J.M. Guinn, *History of the State of California and Biographical Record of the San Joaquin Valley*, (Chicago, Illinois: The Chapman Publishing Co., 1905).

27. Graves, "California Memories."

28. *Reno Evening Gazette*, 4 April 1901. Sparks already owned 10,000 acres in Texas and his will does not mention the additional 12,000 acres.

29. In 1903 Elko County tax assessment rolls listed Sparks-Harrell as the third-largest taxpayer in Elko County with an assessed value of $242,425, *Wells State Herald*, 18 September 1903.

30. *Harper's Weekly*, Vol. XLVII (1903):1017–34.

31. Little is known about John Sparks's financial status in Nevada when he died except the general knowledge that he was bankrupt. His estate in Texas listed assets of $47,015 and debts of $57,784 in Probate No. 1554, Williamson County, Texas, September 28, 1908.

32. Guinn, "History of State of California."

33. For late nineteenth-century scientists who were active in describing the forage resources of the sagebrush/grasslands see F.G. Renner, *A Selected Bibliography on Management of Western Ranges, Livestock, and Wildlife,"* (Washington, D.C.: GPO, Misc. Publ. 281, U. S. Dept. of Agric., 1938).

34. *Wells State Herald*, 15 August 1905.

35. Aven Nelson, *The Red Desert of Wyoming*, (Washington, D.C.: GPO, Bull. No. 13, Div. of Agrostology, U. S. Dept. of Agric., 1898).

36. John Sparks purchased the Wedekind mine for $175,000 in cash (a newspaper editor saw the money). Residents of Reno thought Wedekind was crazy because he kept prospecting in the ridge north of the city, see *Reno Evening Gazette*, 31 May 1902. It is remarkable that John Sparks, who had invested in mining ventures in Wyoming in the 1870s and in Nevada for twenty-five years, could have been victimized in the Wedekind mine fraud. Apparently, the surface deposits were rich oxides of silver and the mill was built to treat such ores; but at depth, the ores became sulfates that the mill was unsuccessful in treating. The mine also encountered a great deal of hot water.

The area is a valid mineral prospect and Sparks probably suffered from bad luck and poor business judgement, see H.F. Bonham, *Geology and Mineral Deposits of Washoe and Storey Counties, Nevada*, (Reno, Nevada: Nevada Bureau of Mines, Mackay School of Mines, Bull. 70, University of Nevada, 1969). After his death, Sparks's ranch in Texas became the site of the Texaco oil fields.

37. John Clay, *My Life on the Range*, (Chicago, Illinois: Privately Printed, 1923).

38. *Water and Related Land Resources*, (Reno, Nevada: Humboldt River Basin Nevada, Basinwide Rpt. No. 12, Max C. Fleischmann College of Agriculture, University of Nevada, 1966).

37. This is the only known photograph of Jasper Harrell. He was often called Jasper "Barley" Harrell because of his habit of draping a sack of barley over his saddle to feed his horse on long trips.

38. In the 1890s, John Sparks was a nationally known figure in the Stockgrower's Association and American Hereford Association. Later, he was governor of Nevada, for whom Sparks, Nevada, is named.

39. Elnora Sparks, John Sparks's second wife, helped him build his Nevada empire.

40. When Elnora Sparks went camping, there was a wood floor in the tent and her husband's picture on the door.

41. John Sparks's son, Leland Sparks (left), is pictured here with an unidentified riding companion.

42. This is a typical hunting camp on one of the Sparks-Harrell ranches. Note the Dutch oven with pot racks and hooks and the chuckwagon in the background. The two boys (right) are Benton and Charles Sparks.

43. John Sparks was a hot-rod buggy driver and raised hounds to run coyotes. One year he wore out four sets of wheels.

44. John Sparks's annual hunting parties attracted prominent political and financial figures; even President Roosevelt enjoyed visiting a Sparks's hunting camp.

45. Even if mule deer were scarce, the Sparks's hunting parties found sage-grouse in abundance.

46. John Sparks (right) typically used his famous rifle "Alcade" on hunting trips. Note the individual on the left carries a very modern rifle for the time, a Savage.

47. John Sparks (3rd from left), and Judge Bartlett (4th from left), are pictured here with a group of unidentified men who, no doubt, helped tame the Nevada desert.

48. Sparks not only left his mark on the cattle business, but later, as Governor Sparks, he meted out justice. Here he listens during the clemency hearings for Indian Johnny's murder conviction.

CHAPTER
13

RESULTS OF THE GRAND EXPERIMENT

Assume we are standing on the Central Pacific tracks in Elko, Nevada, on a late autumn day in 1900. South of town, a switch engine huffs and puffs as it spots cattle cars at the loading chute in the stockyards. A flashily dressed cattle buyer chews on a cigar and congratulates a sun-and-wind weathered rancher on the quality of the steers he is shipping this fall. The steers have the white faces of Herefords, and they are massive animals compared to the Longhorns driven up from Texas three decades before. The rancher wonders if the check in his pocket is going to satisfy the banker when they meet next morning, and worries whether it will provide adequate working capital to keep him going until the next seasonal marketing. Pride in what he has built, and a bittersweet love affair with the cold desert have kept him trying to make it on the sagebrush/grasslands.

The cowhands working the corrals are dead tired from the five-day drive from home range to Elko stockyards. When the last steer is loaded, it will be payday and time for a spree. They do not hurry the big four-year-old steers, but keep steady pressure on the animals in the crush-corral leading to the chute. They practice an intricate craft of manipulating mostly wild animals at close quarters.

A huge, empty freight wagon rumbles across a grade crossing as the teamster wheels an eight-horse hitch to back the wagon to a loading dock on Front Street. The mercantile stores at Elko supply and pack a multitude of supplies, from thimbles to coal oil, to help keep isolated ranch headquarters operating through the winter months when roads are impassable. Freighters are anxious to finish loading and get underway, as there is the bite of frost in the air, and a sharp wind hints an early winter.

The scent of burning juniper wood drifts down from the residential areas of Elko, where the wives were up at dawn, stoking kitchen stoves to fry ham, eggs, and hot cakes for husbands and children. Lunch pails are packed; brakemen and conductors are off to the rail yards; clerks are opening stores and banks.

But the commercial appearance of downtown Elko has something missing. There are no spring wagons of farmers' families loaded with sun-bonneted wives and scrubbed children. Tillers of the soil are not a part of the sagebrush environment of Nevada at the turn of the century. But on the streets we see broken-down wagons from starvation ranches, and pack strings loaded with supplies for small migratory sheep outfits, with total capital of 500 old ewes. Dirt farms and farmers are very scarce.

A gaunt, booted cowboy walks sheepishly out of Adobe Alley on the wrong side of the tracks, pauses to roll a smoke before starting uptown for a much needed cup of coffee or some of the "hair of the dog." He searches his pockets for a coin for the old Indian panhandling at the depot corner. The cowboy is somewhere between the acceptable and unacceptable divisions of Elko society. At times, both sides of the tracks envy his cowboy lifestyle, but be belongs to neither.

A freight train works up steam to get underway in the rail yards east of town. Brakeman and conductors wave signals while the sweating fireman shovels coal into the firebox. And as the train gathers speed, shadowy figures in faded blue denims and worn work shirts emerge from the willows beyond the stock corrals. They wear slouch hats, and have slack bedrolls tied over one shoulder. These are the seasonal hay hands and hobos, waiting to slip aboard the west-bound freight, in order to escape the cold of a high-desert winter, where there is no demand for hands once the haying is finished.

That was the picture in Elko, Nevada as the twentieth century unfolded. The turn of the century became a watershed division between the previous turbulent years of the "Grand Experiment" in ranching, and the following period of uncertainty that was to continue until drought, depression, and war brought dramatic change near mid-century. The significant period of experimentation in the sagebrush/grasslands had reached its conclusion by 1900.

The "Grand Experiment" had brutally disclosed some limitations of the sagebrush/grasslands environment. Forage had to be conserved for wintering of the next livestock. Forage could only be produced through irrigation. Water was a finite natural resource. Individual ranches had extended their capital and engineering potential to the limit in developing irrigation systems. Future development would require cooperative capital or, unheard of at the time, federal funds for development of irrigation.

Ranchers had potential to control finite water rights at state level through legislative action, but there was lack of common purpose to accomplish this. The finite nature of water resources in Nevada probably contributed to this temporary lack of resolve, because any distribution of water rights based on available stream flow would result in some claimants coming up dry.

Ranching in the cold desert seemed to breed contrary independence among practitioners. Ranchers did not have the right to legislative action to control grazing on public lands because those lands belonged to the federal government. The agrarian society of sagebrush ranchers could not obtain sufficient land to economically support a livestock operation. Failure to resolve this problem burdened and threatened to destroy the range livestock industry of the sagebrush/grasslands.

Increasing popularity of rangeland sheep operations increased competition for grazing resources. Side effects of such conflicts were the dedication of Forest Reserves and eventually National Forest where there were virtually no trees. Support from established ranches was based on the proposal by the federal government that grazing on the proposed National Forest be based on historic use and the ownership of commensurate property capable of supporting livestock for the portion of the year they were not being grazed on the National Forest ranges.

With the passing of generations, the various ethnic groups that made up the initial ranch culture tended to blend and assume a native quality. Most notable exceptions were the American Indian and the Black cowboy. The Indian was largely outside the mainstream of political, economic, and social life. The Black ranchhand would soon virtually disappear from the ranching scene in the sagebrush/grasslands.

The stock-raising society was moving, flowing, struggling each strata on its own way toward economic, social, and environmental equilibrium in the cold desert.

EQUIVALENCY NAME TABLE

Plants

Abies lasiocarpa—subalpine fir
Agoseris glauca—pale agoseris
Agropyron smithii—western wheatgrass
Agropyron spicatum—bluebunch wheatgrass
Agrostis alba—redtop
Artemisia arbuscula—low sagebrush
Artemisia cana—silver sagebrush
Artemisia nova—black sagebrush
Artemisia spinescens—budsage
Artemisia tridentata—big sagebrush
Artemisia tridentata ssp. *tridentata*—basin big sagebrush
Artemisia tridentata ssp. *vaseyana*—mountain big sagebrush
Artemisia tridentata ssp. *vaseyana* form *speciformis*—subalpine big sagebrush
Artemisia tridentata ssp. *Wyomingenis*—Wyoming big sagebrush
Atriplex—saltbush
Atriplex canescens—fourwing saltbush
Atriplex confertifolia—shadscale
Balsamorhiza sagittata—arrowleaf balsam root
Bromus tectorum—cheatgrass, downy brome
Carex sp.—tuff sod of sedges
Castilleja sp.—Indian paintbrush
Ceratoides lanata—winterfat
Cercocarpus ledifolius—curlleaf mountain mahogany
Chrysothamnus viscidiflorus—low rabbitbrush
Collinsia sp.—Chinese house
Cowania mexicana ssp. *stansburiana*—cliffrose
Dactylis glomerata—orchard grass
Delphinium sp.—low larkspur
Delphinium barbeyi—tall larkspur
Delphinium glaucum—tall larkspur
Elocharis aricularis—spike rush
Elymus cinereus—Great Basin wildrye
Elymus triticoides—diminutive creeping wildrye
Ephedra viridis—green ephedra
Festuca idahoensis—Idaho fescue
Gossypium—cotton
Hordeum brachyantherum—meadow barley
Hordeum vulgare—barley
Iris missouriensis—Rocky Mountain iris
Juncus balticus—wiregrass
Juniperus osteosperma—Utah juniper
Medicago sativa—alfalfa
Mimulus nanus—skunk monkey flower
Muhlenbergia brachyantherum—mat muhly
Oryzopsis hymenoides—Indian ricegrass
Phleum pratense—timothy
Pinus albicaulis—white bark pine
Pinus contorta—lodgepole pine
Pinus flexilis—limber pine
Pinus longalva—ancient bristlecone
Pinus monophylla—single-leaf pinyon
Poa nevadensis—Nevada bluegrass
Poa sandbergii—Sandberg bluegrass

Populus fremontii—Fremont cotton-
wood
Populus nigar var. *italica*—Lom-
bardy poplar
Populus tremuloides—quaking aspen
Populus trichocarpa—black cotton-
wood
Prunus andersonii—desert peach
Purshia tridentata—bitterbrush
Ribes velutinum—Ribes
Salix—willow
Sarcobatus vermiculatus—greasewood
Scirpus robustus—alkali bullrush
Sitanion hystrix—squirreltail
Stipa comata—needle and thread
grass
Stipa thurberiana—Thurber needle-
grass
Symphoricarpos spp.—Snowberry
Tetradymia canescens—horsebrush
Typha latifolia—cattail tule
Vulpia octoflora—six-weeks fescue
Wyethia mollis—mules ears
Zigadenus venenosus—death camas

Animals

Antilocapra americana—pronghorn
Bison bison—American bison
Bos—cow
Camelops—camel
Castor canadensis—beaver
Centrocerus urophasianus—sagegrouse
Cervus canadensis—elk, wapiti
Eguus caballus—horse
Gymorhinus cyanocephala—pinyon
jays
Lepus californicus—black-tailed jack-
rabbit
Lepus townsendii—white-tailed jack-
rabbit
Mastodon—Mastodon
Odocoileus hemionus—mule deer
Ovis aires—sheep
Ovis canadensis—desert bighorn
Pogonomyrmex—harvester ants

Bacteria

Bacillus tularense—tularemia

INDEX

ABOUT THE AUTHORS

James A. Young had the good fortune to be born and reared in a ranching environment where it was impossible to escape the influence of pioneer cattlemen. In a box beneath his desk are ranch ledgers prepared by his father. They reach back nearly a century to the ranches of his great-grandfather who raised beef to feed the goldminers. A distinguished research scientist, Dr. Young has spent over twenty years in Nevada devising methods to maintain and preserve the soils and plant communities that constitute the rangelands in the Great Basin. He feels that a vital factor in the conservation of rangeland resources is how the general public perceives the resource. Writing to provide a historical perspective on the use of rangelands is a major interest of Dr. Young's and a way to understand both past and present with an eye to future usage.

B. Abbott Sparks comes from a New Lands pioneer family that moved westward from the South, to Texas and Oklahoma, and ultimately, the West. He is also the great-nephew of John Sparks, whose life is chronicled in this volume. Following graduation from Oklahoma University, Abbott Sparks entered the publishing field. He developed an international organization, Petroleum Engineer Publishing Company, which circulated magazines, newsletters, and books throughout 115 countries. This company was later sold to Harcourt Brace Jovanovich. After thirty-five years in publishing, Sparks still maintains a small international organization, but is devoting his later years to his family and to a list of ongoing literary projects that cover a broad spectrum ranging from his fascination with the western experience to his love of travel.

A NOTE ON THE TYPE

The text of this book was set in a digitized version of Goudy Old Style, a modern classic type design. The typeface was named for Frederic W. Goudy, a self-taught type designer who gained international recognition in the early 1900s. After an early professional life marked by failure and financial losses in several vocations, Goudy designed his first typeface in 1896 at the age of thirty. He ultimately created 116 type designs—the last one in 1944 at the age of seventy-nine. His early typefaces were considered good designs, but it was not until after a fire devastated his work in 1908 that he began to develop type styles that reflected his genius as a designer. Goudy Old Style, originally cut in 1915, is noted for its simplicity, dignity, and strength.